CHEMISTRY RESEARCH AND APPLICATIONS

PARAMAGNETIC COMPLEXES IN HIGH RESOLUTION NMR SPECTROSCOPY

Chemistry Research and Applications

Additional books in this series can be found on Nova's website under the Series tab.

Additional e-books in this series can be found on Nova's website under the e-book tab.

CHEMISTRY RESEARCH AND APPLICATIONS

PARAMAGNETIC COMPLEXES IN HIGH RESOLUTION NMR SPECTROSCOPY

V. K. VORONOV[1]
AND
A. V. PODOPLELOV[2]
[1]Ikutsk State Technical University (Irkutsk, Russia)
[2]The company HTLab AG (Pfaffikon, Switzerland)

nova publishers
New York

Additional color graphics may be available in the e-book version of this book.

Library of Congress Cataloging-in-Publication Data
Paramagnetic complexes in high resolution NMR spectroscopy / V.K. Voronov and A.V. Podoplelov (Irkutsk State Technical University, Russia, and others), editors.
pages cm. -- (Chemistry research and applications)
Includes bibliographical references and index.
ISBN 978-1-63482-698-3 (hardcover)
1. Nuclear magnetic resonance spectroscopy. 2. Electron paramagnetic resonance spectroscopy. 3. Spectrum analysis. 4. Chemistry, Organic. I. Voronov, V. K. (Vladimir Kirillovich) II. Podoplelov, A. V.
QD96.N8P37 2015
543'.66--dc23

2015017728

Published by Nova Science Publishers, Inc. † New York

CONTENTS

PREFACE

Just after the discovery of the nuclear magnetic resonance (NMR) phenomenon in the condensed media, it became clear that it will constitute a basis of a powerful method for investigation of structure and properties of molecules. This conclusion has been confirmed by numerous publications devoted to the application of NMR in diverse research areas. High efficacy of the NMR in the solution of various problems related to study of structure and behavior of multielectron (molecular) systems is due, first of all, to the progress in NMR spectrometers and to achievements in understanding this phenomenon. The latter is supported by a section of molecular spectroscopy which is based on the analysis of NMR spectra transformed upon complex formation. Primarily, we mean here the complexes, the specifics of which is defined by electrons located on $\boldsymbol{d}-$ and $\boldsymbol{f}-$orbitals. In particular, it has been shown that the studying of NMR phenomenon in paramagnetic complexes allows valuable information on structure of molecules to be obtained. This possibility follows from peculiarities of the NMR spectra of paramagnetic systems.

The electron-nuclear or hyperfine coupling (HFC) between unpaired electrons and resonant nuclei induces characteristic shifts and broadening of signals, which are a source of information about electron and spatial structure of ligands in paramagnetic complexes. The fundamentals of NMR theory in paramagnetic systems were formulated by Blombergen, McConnell, Solomon, Berkheyn and other researchers. However, wide application of HFC specifics for the solution of diverse chemical tasks was triggered later, after development of a method for analysis of NMR spectra transformed by additives of paramagnetic complexes. Apparently, the pioneering work (in which one of authors of the present monograph was engaged) in this line was performed under the supervision of academician Yu.N. Molin. The results of this work were published in 1968. One should also mention the pioneering works of Hinckley (1969), Sanders and Williams (1970) which gave impetus to intensive investigations dealing with use of paramagnetic complexes of rare-earth elements with organic ligands in the NMR spectroscopy. As a consequence, a method, allowing structural information for liquids to be obtained at molecular level, has been elaborated. It should be noted that informational content of this method is comparable to X-ray diffraction analysis of solids.

The monograph covers theoretical aspects of the method based on the analysis of NMR spectra transformed by electron-nuclear or hyperfine coupling and provides examples of its practical application for the solution of different tasks related to structure and intramolecular dynamics of multielectron systems. The book consists of three chapters. Chapter 1 discloses the fundamentals of high resolution NMR spectra transformed by hyperfine coupling. Two other chapters illustrate fruitfulness and prospects of the application of NMR phenomenon in paramagnetic systems for studying molecular structure. Examples of specific spectra, analyzed in the monograph, are selected in such a way that they could help a reader to understand a circle of possible problems which are solved using peculiarities of the NMR phenomenon in paramagnetic systems. The same examples illustrate also the techniques which can be employed for obtaining the information on molecular structure of both ligands and complexes using their NMR spectra transformed by hyperfine coupling.

Here, it is pertinent to note another characteristic feature of the NMR spectra transformed by hyperfine coupling. Very often the application of paramagnetic shifting reagents allows to simplify a shape of the spectra recorded on spectrometers operating at proton frequencies of not higher than 100 MHz. In the result, the spectra can be analyzed according to the rules used for the spectra of first order. Therefore, there is no need for expensive NMR spectrometers having much higher resonant frequencies.

Despite the significant progress in studying of the NMR peculiarities in paramagnetic compounds, a number of unresolved problems in this field of knowledge still remain. Obviously, this is due to the fact that the studies of HFC with a help of NMR method are still sparse because of a limited number of paramagnetic systems, where electron-nuclear interaction can be investigated using high resolution NMR spectra. Therefore, one can state that the whole classes of the paramagnetic complexes containing diverse heteroatom compounds as ligands "wait for the turn" for their HFC study. Such studies should be favored by ever-increasing research interest in metal complexes with radical and redox-active ligands. We can emphasize here that modern NMR spectrometers with resonant proton frequencies of 400 MHz, 500 MHz and higher, offering ample possibilities for manipulation with the detected spectral information, allow recording the spectra of paramagnetic systems in wide spectral ranges (about 400 ppm for ^{1}H NMR and 2000 ppm for ^{13}C NMR). Therefore, one should expect that a number of tasks (including fundamental ones), which solution is connected with the detection of HFC spectral manifestation, will increase dramatically. In particular, the above conclusion was illustrated by the NMR studies in complex sterically unhindered paramagnetic systems containing multielectron heteroatom structures as surroundings of the central metal (Sagdeev R.Z., Voronov V.K., Podoplelov A.V. et al. Peculiarities of the NMR spectra of heterospin complexes // Izv. Akad.Nauk. Ser. Khim. 2001, No. 11, p. 1986 – 1994).

Just after first reports on application of paramagnetic additives in structural studies, two monographs written by known scientists have been published [1, 2]. The first one deals with the general HFC theory in paramagnetic systems as well as with the effect of this interaction on parameters of the NMR spectra. In the second monograph, the problems related to use of lanthanide shifting reagents for interpretation of NMR spectra of heteroatom organic compounds, are discussed. After appearance of the abovementioned monographs, by the end of the eightieths of the twentieth century, almost all so far known papers on paramagnetic complexes were published. Also, the investigations into a mechanism of uncompensated electron spin delocalization and some other aspects of HFC were carried out. The results obtained were summarized in monograph [3] published in 1989. This monograph is at the heart of the present book which was revised taking into account the results obtained over the last years.

Chapter 1

Application of NMR Spectra of Paramagnetic Systems for Structural Studies of Organic Ligands

In spite of the fact that the first work on NMR in paramagnetic systems was published in 1949 [4], prior to the beginning of the sixties of the twentieth century, the investigations in this area of NMR spectroscopy were few in number. First of all, this is due to the lack of suitable systems, where high-resolution spectra can be detected. The breakthrough in this direction was made by Eaton et al. In the NMR spectra of the paramagnetic complexes they managed to find that width of the resonant lines did not exceed some hertz [5]. However, the results obtained by this research team could not reflect the specifics of HFC observed in the majority of known heteroatom organic compounds.

As a rule, one fails in recording the high resolution NMR spectra of paramagnetic complexes because of significant broadening of the signals. This relates, first of all, to the NMR signals from carbon, nitrogen and other atoms which constitute skeleton of the molecule. For example, signals broadening from protons can reach hundreds and even thousands hertz. The carbon signals can broaden so significantly that their detection becomes impossible at all. The values of paramagnetic shifts can quite high (in comparison with the typical spectral range of diamagnetic compounds). Besides, the paramagnetic shifts can be different in sign that complicates detecting of signals related to individual paramagnetic particles (complexes and radicals). However, it turns out that these difficulties can be circumvented if to employ samples containing considerable excess of ligands as compared to concentration of the coordination ion. Then the spectrum becomes average and depends, in particular, upon character of hyperfine coupling in a paramagnetic complex. For a long time (up to the beginning of the ninetieths of the last century), overwhelming number of works connected with usage of HFC specifics for studying molecular structure were carried out in such a way.

The increase of a number of compounds, both synthesized and isolated from natural feedstock, significantly expands a gamut of products appropriate for HFC studies. Also, the situation began to change with the appearance of commercial Furrier-spectrometers working in a pulse mode. The application of modern NMR spectrometers allows recording the spectra of paramagnetic systems in a wide spectral range (about 400 ppm for ^{1}H NMR and 2000 ppm for ^{13}C NMR). Ultra-speed work stations of such spectrometers are capable of processing a huge number of 2M-spectra that makes it possible (at least, in principle) to assign the resonant signals. The usage of temperature attachments enables to perform complex dynamic investigations involving paramagnetic systems. Therefore, we are sure that studies of paramagnetic compounds using high resolution NMR spectroscopy will be progressing intensively.

This chapter summarizes the results of fundamental investigations aimed at elucidation of the hyperfine coupling nature as well as spatial and electronic structure of organic heteroatom compounds. The authors did not try to cover all the relevant papers, but intended to pay the attention to all existing ideas concerning application of paramagnetic complexes for structural studies of organic ligands. Starting from this assumption, the most general works with the corresponding references to other publications are discussed. Structural studies of paramagnetic shifts induced by organic radicals upon the formation of hydrogen bonding with organic compounds are beyond the scope of this chapter (see, for example, [6]).

1.1. Nature of Hyperfine Contact Coupling in Paramagnetic Complexes

The specifics of NMR phenomenon in paramagnetic systems is defined by a character of hyperfine coupling between unpaired electrons and resonance nuclei that leads to broadening and shifts (contact and pseudo-contact) of the resonance lines. The contact shift is observed when a probability of the unpaired electron disposition in the location of resonance nucleus is different from zero. Therefore, these shifts can be considered as a source of information about nature of the metal–ligand bond as well as about electron structure of the ligands. The excellent examples illustrating efficiency of the contact HFC application for structural study of multielectron systems are given in the review [5]. The pseudo-contact shift is observed when the dipole-dipole coupling between magnetic moments of the unpaired electrons and nuclei is not averaged to zero. The pseudo-contact shifts characterize spatial structure of the molecule that explains their use for the solution of different structural problems. The most striking example of successful application of pseudo-contact shifts for these purposes is a method of lanthanide shifting reagents offered in the works [7, 8]. These works have initiated a large number of publications (especially in the first half of the 70th of the twentieth century), including reviews (more than 40). The current notions on the HFC mechanism in paramagnetic complexes of first transition group rare-earth elements are summarized.

1.1.1. Complexes of First Transition Group Elements

The majority of the published works deal with the ^{1}H NMR study of delocalization of unpaired spin density from the coordination ion to ligands and across ligands in nickel and cobalt complexes with heterocyclic compounds. Interpretation of the experimental paramagnetic shifts in the NMR spectra is based on the ideas formulated in the early sixties of the XX century [9, 10]. It is established that the perturbation effect in unpaired electrons is transferred to resonance nuclei of the ligands across both $\pi-$ and $\sigma-$ bonds ($\pi-$ and $\sigma-$ delocalization mechanisms). It appears that delocalization across the $\sigma-$bonds of unsaturated molecules of the ligands, upon removal from the coordination ion, leads to fast attenuation of the induced unpaired spin density. In this case, paramagnetic shifts have the same signs. If the unpaired spin density is located in $\pi-$system, high values of paramagnetic shifts are detected for the signals of nuclei separated from the central ion by several bonds (long-range delocalization). Besides, signs of shifts become alternating. In the spectra of saturated compounds, shift values can also have alternating signs, and electron-nuclear coupling is observed for nuclei which are considerably remote from the coordination ion. These experimental data were rationalized by the transfer of perturbation effect of electron spin not only across the bonds (spin delocalization), but also according to mechanism of spin polarization.

The progress in NMR spectrometers makes it possible to investigate paramagnetic complexes using high resolution NMR spectra recorded from other nuclei (nor protons only). Advances in quantum-chemical calculations also favor such studies [11]. Paramagnetic shifts in ^{13}C NMR spectra are of special interest. The pioneering works in this field have been published in 1970. They describe paramagnetic systems, first, with well-established nature of proton-electron coupling and, second, containing ligands, the molecules of which represent typical $\sigma-$ or $\pi-$ systems. Representative example of such studies is the paper [12], where the mechanism of hyperfine contact coupling in a complex (I) of $N,N'-$di($p-$tolyl)aminotroponoiminate coordinated with nickel is discussed on the basis of ^{13}C NMR spectra.

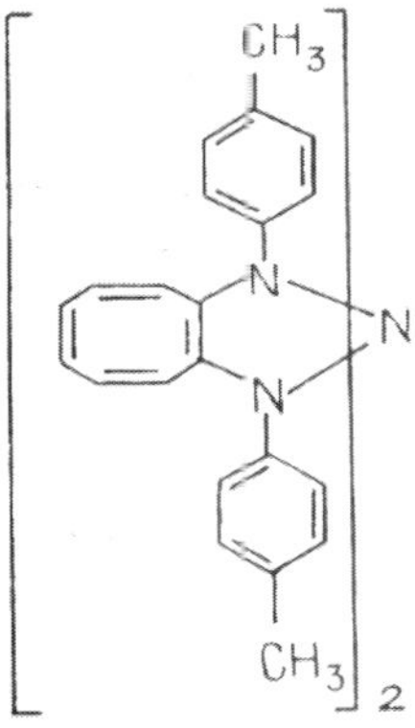

I

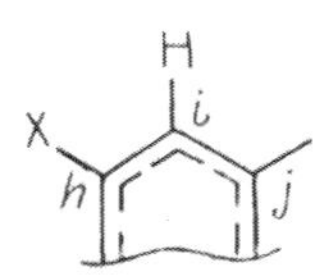

Ia

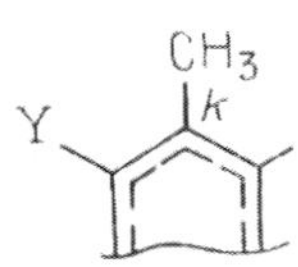

Iб

It is found that contact shifts are caused by the uncompensated electron spin density which is only in $\pi-$system. Therefore, taking into account the results given in works [13, 14], the expression for fragment Ia is as follows:

$$\omega_C^i/\omega_H^i \cong 2,5\left(\rho_C^h+\rho_C^j\right)/\rho_C^i, \tag{1.1}$$

and for fragment Ib is

$$\omega_C^{CH_3}/\omega_H^{CH_3} \cong 2,1, \tag{1.2}$$

where ρ – spin density on p^{π} – orbital of the corresponding carbon atom; ω - experimental values of contact shifts.

The ratios ω from the expressions (1.1) and (1.2) are evaluated using literature values ω_H and ρ [9]. It is found that the calculated and experimental ω_C/ω_H values are in bad agreement. At the same time, the calculated $\omega_C^{CH_3}/\omega_H^{CH_3}$ value differs slightly from the experimental values. The authors of the work [12] rationalize this finding by the fact that unpaired spin density is out of the σ – system of the ligand molecule due to $\pi-\sigma$ polarization mechanism. The performed investigations allow them to infer that the spin densities, calculated from proton contact shifts, are probably unreliable. This conclusion is supported by the data given in [15]. As a consequence, it is postulated that ^{13}C nuclei are more sensitive (in comparison with ^{1}H) to contact coupling and, hence, more adequately reflect the mechanism of such coupling transfer. However, the study of isotropic contact shifts of ^{13}C signals in paramagnetic complexes of nickel with organic ligands reveals a linear ratio between uncompensated electronic density on carbon and hydrogen atoms in the same position [16]. This important inference permits determining in some cases the parameters of hyperfine coupling for ^{13}C nuclei from the corresponding measurements of proton spectra.

Limited potential of quantum-chemical calculations in determination of electron structure of multielectron systems actualizes the approaches modeling the delocalization of unpaired spin density and allowing theoretical evaluation of electron-nuclear coupling parameters. Apparently, one of the first physically reasonable models of unpaired spin density delocalization in paramagnetic complexes was offered in [17]. The method was based on comparison of hyperfine coupling constants in the radical **A**, in fragment of the paramagnetic complex **B** and spin-spin coupling constant J_{HH} in the diamagnetic molecule **C**.

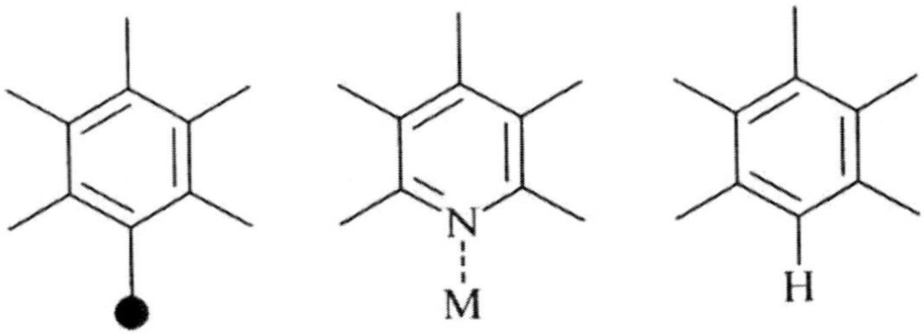

A **B** **C**

Mechanisms of the following perturbation effects were discussed:

1) N nucleus on other nuclei in a diamagnetic molecule through carbon atom;
2) Unpaired electron located on the same carbon atom;
3) Unpaired spin density of nitrogen atom, where the density is transferred from coordination ion.

It was shown that the perturbation effect of unpaired electron on ligands in a paramagnetic complex qualitatively corresponds to that of the electron spin in the radical and nucleus spin in a diamagnetic molecule, which is isostructural relative to the radical and ligand molecules. On this basis, the authors of work [18] have performed the detailed analysis of the contact coupling mechanism in $\sigma-$ systems (on the example of pyridine and its methyl derivatives). The investigations involve the comparison of experimental values ω of paramagnetic shifts and constants of hyperfine coupling calculated by the INDO method for radicals, isostructural towards the ligand molecule (phenyl, $o-$, $m-$ and $p-$ tolyl radicals). The good agreement between theoretical and experimental values is observed. This evidences that paramagnetic shifts in ^{1}H and ^{13}C signals of the pyridine bases in octahedral nickel complexes are induced due to delocalization of uncompensated electronic density only across $\sigma-$ system of the ligand molecule. Moreover, the correct explanation of values and signs of the paramagnetic shifts allows drawing the following conclusions [18].

1. The low-field signals shift and quite fast decreases of absolute values ω upon removal from the coordination center is observed for protons which are directly coupled with the pyridine ring carbons and is not mandatory for methyl protons.
2. The alternation of paramagnetic shift signs in ^{13}C signals is predicted by the calculations. Therefore, it cannot to be considered (as it is accepted) as characteristic feature of $\pi-$ spin delocalization. Besides, the calculations correctly predict the shift signs of the methyl group protons. Consequently, the opposite shift of signals assigned to the ring hydrogen and methyl moiety, coupled with the same carbon atom, should not also be interpreted as the proof of the $\pi-$ spin mechanism realization.

A vivid example of successful application of the INDO method for Fermi-interaction transfer across the $\sigma-$ bonds only is given in the work [18], where paramagnetic shifts in the PMR spectrum of 4-vinylpyridine induced by $\mathbf{Ni(AA)_2}$ are studied. The authors compare their analysis of experimental values of contact proton shifts with that of previously performed using the expanded Hueckel approximation method [19]. The INDO method turns out to correctly predict all signs and ω value size in a reasonable semi-quantitative agreement with experiment, while the Hueckel method assuming only the $\sigma-$ mechanism of delocalization cannot explain all the experimental data. The assumption that negative spin density is distributed over the $\pi-$ system of the ligand does not give insight into $\gamma-$ proton shifts, and incorrectly predicts signs of paramagnetic shifts in the signals of $\beta-$ protons of 4-vinylpyridine vinyl group. All the above stated allow concluding [18] that in principle the INDO can be considered as perspective method for interpretation of contact paramagnetic shifts in ^{1}H and ^{13}C signals. After the critical remarks made in papers [20] and [21] (in the latter not phenyl, but cationic pyridine radical is used as a model), Horrocks has analyzed the spin delocalization across $\sigma-$ bonds of the ligand molecule and performed quantum-chemical calculation (INDO) of electron-nuclear coupling constants in paramagnetic complexes of transition group elements [22].

The authors of work [23] have studied proton paramagnetic shifts in the spectra of some complexes of nickel and cobalt tris(acetylacetonates). Critical analysis of quantum-chemical evaluation of the experimental data and limitations of INDO and CNDO calculations is given. On the basis of paramagnetic shifts for $Ni(AA)_2(Py)_2$; $Ni(AA_3^-)$ and $Co(AA)_2(Py)_2$ complexes, it is concluded that the separation method of contact and pseudo-contact contribution to paramagnetic shifts for the cobalt complex is indeed applicable to the axially located ligands (L) in $M(AA)_2 \cdot L_2$ $(M = Ni,\ Co)$ complexes, but not to (AA) complexes. Additionally, in the work [23], the comparability of donor ability of acetylacetone and acetylacetonate is explained, and the inference (made earlier in [24]) that in the octahedral complexes Co^{2+} the third unpaired electron occupies $\boldsymbol{d}_{xy}$ orbital is confirmed. Such point of view does not correspond, according to [23], to the idea about mechanism of spin delocalization on the protons (AA), offered in [25] for explanation of significantly different paramagnetic shifts of these protons in nickel and cobalt complexes. In reality, the problem is even more complex because of the difficulties related to selection of the symmetry type for central ion surroundings. Horrocks [26], using his efficient perturbation method in approximation of the ligands low-field [27], has found that in the nickel complex unpaired electrons occupy $\boldsymbol{d}_{x^2}$ – и $\boldsymbol{d}_{xy}$ –orbitals, and in the cobalt complex are located on $\boldsymbol{d}_{x^2-y^2}$ –, $\boldsymbol{d}_{x^2}$ –, $\boldsymbol{d}_{xy}$ –orbitals. This enables to explain not only the nature of paramagnetic shifts, but also temperature dependence of magnetic susceptibility as well as optical spectra of the studied complexes.

^1H and ^{13}C NMR study of the effect of ligand and chelate nature on covalence degree of the metal–ligand bond in nickel complexes has been carried out [28, 29]. Similar results are reported in [30]. The map of spin density distribution over all atoms of a ligand (pyridine and its methyl derivatives) is compiled [31, 32] to rationalize the delocalization mechanism in paramagnetic complexes. It is concluded that correlation between HFC constants in paramagnetic complexes and spin-spin coupling constants in diamagnetic molecules (isostructural relative to the ligand molecules) is no general and is valid only for light nuclei. Even earlier, the unsatisfactory correlation between values of paramagnetic shifts in signals of pyrazole and imidazole protons and constants of the pyrrole molecule has been noted [33].

The nature of contact hyperfine coupling in ML_4T_2 and ML_2T_2 complexes ($M = Ni^{2+},\ Co^{2+},\ L-$pyridine and alkylpyridines, $T-CF_3CO_2$) was discussed [34]. It was assumed that the value of pseudo-contact shift induced by nickel paramagnetic ion was insignificant that allowed defining the contact component from cobalt signals shift. Also, it was supposed that mechanisms of uncompensated electron spin delocalization in complexes $\boldsymbol{Ni^{2+}}$ and $\boldsymbol{Co^{2+}}$ were similar. Experimental ω^{Ni} values and calculated ω^{Co} values were used for discussion of the mechanism of uncompensated electron spin delocalization across the ligand molecule on the basis of a quantum mechanical evaluation of spin-spin coupling constants. The spectra of individual complexes solution were recorded that allowed establishing the peculiarities of their stoichiometry at different temperatures [34].

To study the nature of HFC in five-membered $\boldsymbol{N}-$heterocycles, paramagnetic shifts ω for protons of 1-vinylimidazole (VIM) coordinated with manganese, cobalt, nickel, and copper chlorides were determined [35]. Experimental ω values were used for calculation of hyperfine coupling constants $\boldsymbol{A}$ according to the following expression:

$$A = \frac{33{,}89 \cdot \omega_k \cdot T}{S(S+1)} \cdot 10^{-6}. \tag{1.3}$$

In this expression, $\boldsymbol{A}$ is defined by the value of uncompensated electron density ρ on the corresponding $\boldsymbol{s}-$orbital, E; ω_k – is contact paramagnetic shift, ppm; $\boldsymbol{T}-$is temperature of the sample; $\boldsymbol{S}-$is total spin of the complex.

Selected results obtained in [35] are shown in Figure 1 – Figure 3. For detailed studying of HFC in the complexes, the above-described modeling of delocalization of uncompensated electron spin density and allowing theoretical evaluation of electron-nuclear coupling parameters has been used. Constants $\boldsymbol{A}$ of protons and nuclei ^{13}C of VIM molecule coordinated with nickel, spin densities ρ on $\boldsymbol{s}-$orbitals of the vinylpyrrole radical and $\boldsymbol{J}_{HH}$ constants in 1-substituted pyrroles were compared. It has been established that $\boldsymbol{\sigma}-$mechanism of delocalization is predominant in the complexes of cobalt, nickel and copper with VIM, whereas the paramagnetic shifts attributed to Mn^{2+} result from HFC transfer across both $\boldsymbol{\sigma}-$ and $\boldsymbol{\pi}-$system of the ligand molecule.

The latter is due to fact that a fragment of VIM molecules in solution coordinates not with pyridine nitrogen atom $(N-3)$, but with pyrrole one $(N-1)$.

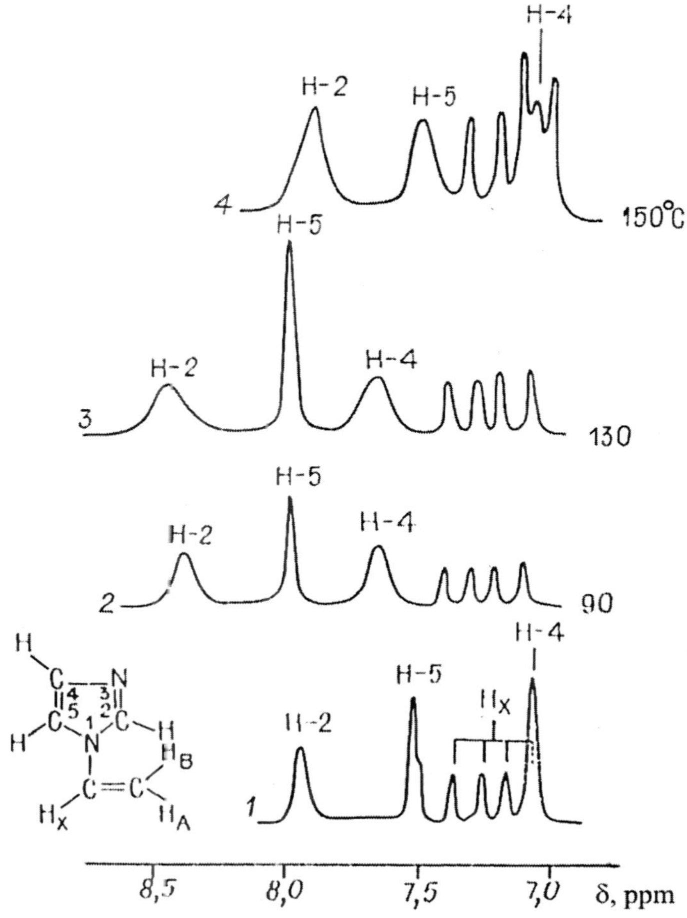

Figure 1.1. Positions of H-2, H-4, H-5 and HX proton signals in PMR spectra of VIM without (1) and with additives of $CoCl_2 \cdot 4VIM$ (2), $NiCl_2 \cdot 4VIM$ (3), $CuCl_2 \cdot 4VIM$ (4).

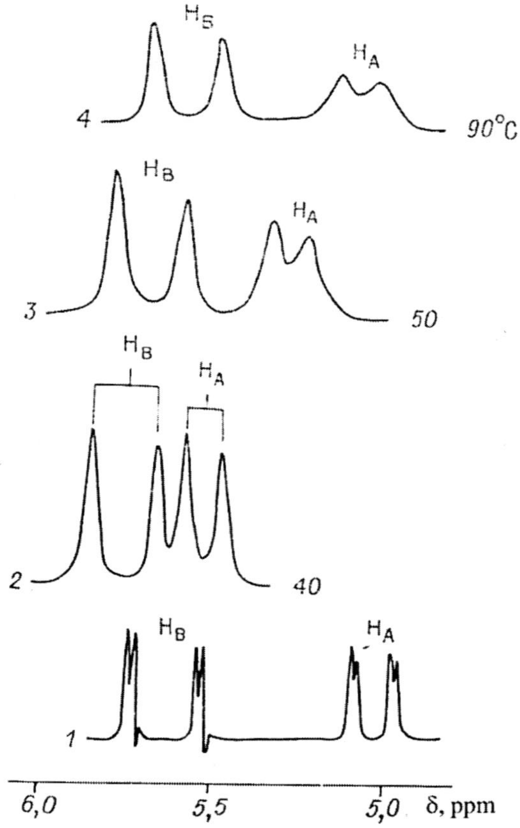

Figure 1.2. Positions of H_A and H_B proton signals of the vinyl group in PMR spectrum of VIM without (1) and with additives of $CoCl_2 \cdot 4VIM$ (2), $NiCl_2 \cdot 4VIM$ (3), $CuCl_2 \cdot 4VIM$.

The ^{1}H NMR study of $Ni(\text{imidazole})_6^{2+}$ and $Co(\text{imidazole})_6^{2+}$ complexes [36] has shown the ratios of contact paramagnetic shifts $(\omega_k)_{4,5}/(\omega_k)_2$ of the imidazole cycle protons are 1,10 and 1,62, respectively. In opinion of the authors [36], this fact contradicts the common literature statement that the delocalization mechanism of uncompensated electron spin density is similar for complexes of cobalt and nickel. However it has been noted [23] that such similarity should be realized for axially coordinated ligands. In Co^{2+} and Ni^{2+} complexes, six imidazole molecules are located both in axial and equatorial positions [36].

It is found [35] that for nickel and cobalt complexes containing four molecules of VIM each, the ratio $\left[\frac{1}{2}(A_4 + A_5)\right]/A_2$ equals 1,22 and 1,04, while the ratio A_4/A_2 (shifts of H-2 and H-4 signals, located in direct proximity to the coordination center, are defined by the dominating HFC mechanism) is 1,40 and 1,14. These values qualitatively (higher value relates to the cobalt complex) and to some extent quantitatively correspond to the similar ratios for Co^{2+} and Ni^{2+} complexes coordinated with six imidazole molecules. It is postulated [23] that due to similarity of electron delocalization mechanisms in cobalt and nickel complexes, the replacement of two axially located imidazole molecules should not essentially affect the relative values of contact paramagnetic shifts. Thus, it is beyond reason to believe that experimental data given in [36] contradict the statement about similarity of delocalization mechanisms of uncompensated electron spin density in the corresponding nickel and cobalt complexes.

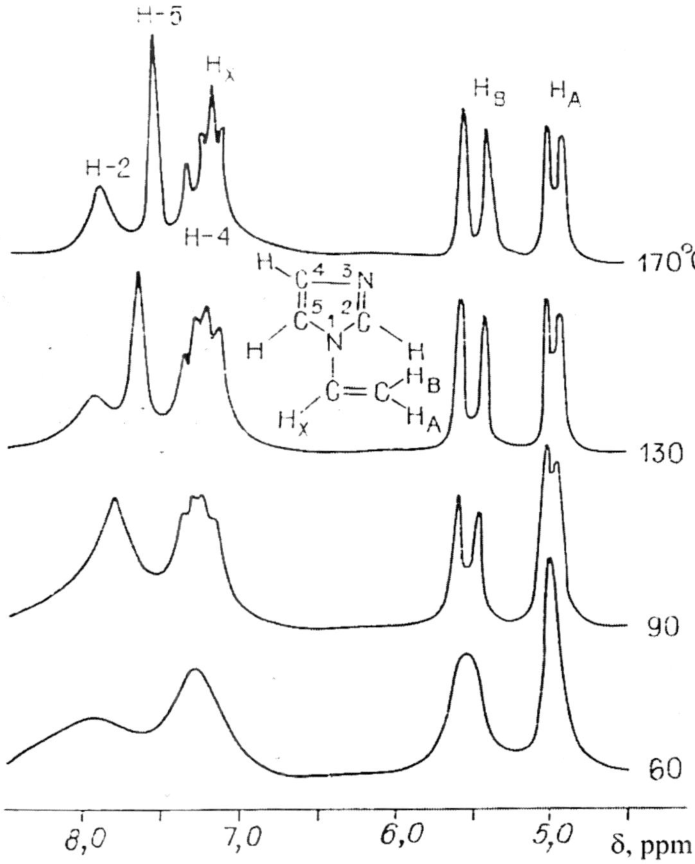

Figure 1.3.PMR spectra of VIM with additive of paramagnetic complex $MnCl_2 \cdot 4VIM$ at different temperatures.

Other researchers also study the NMR spectra of contact interaction [37 - 41].

As for methodological aspects of the problem, the results of the NMR study (mainly ^1H and ^{13}C) of contact hyperfine coupling can be summarized as follows:

1) Successful interpretation of the paramagnetic system spectra and elucidation of the hyperfine contact interaction mechanism depend upon possibility of the performance of correct quantum-chemical calculations of electron structure of the compounds studied;
2) The study of hyperfine coupling without application of quantum-chemical calculations may yield valuable results (especially for new paramagnetic complexes). The success of such studies is defined by the selection of the corresponding model systems.

1.1.2. Complexes of Rare-Earth Elements

Unlike the complexes of first transition group elements, the nature of electron-nuclear coupling in paramagnetic complexes of rare-earth elements with organic ligands was studied poorly before the development of the lanthanide shifting reagents method (LSR). However due to wide application of lanthanide diketonates, the need for such studies became obvious. The employment of LSR in structural research is based on the ideas about dipole-dipole nature of hyperfine coupling between uncompensated electron spins and resonance nuclei. Such ideas were formulated in [42] where it was reported on comparison of the PMR and ESR spectra of paramagnetic acetylacetonates of transition elements. But in later publications, it was stated that the contact mechanism can also be used for the explanation of values and signs of paramagnetic shifts. For example, it is deemed that abnormal paramagnetic shifts in ^{1}H, ^{13}C, ^{19}F, ^{14}N, ^{31}P NMR spectra, induced by LSR, are caused by the contact electron-nuclear coulping [43–45]. This mechanism rationalizes the appearance of contact shifts, which values are defined by Blombergen-McConnell expression [5]:

$$\omega_k = -A\langle S_z\rangle / g_N \beta_N, \tag{1.4}$$

where $A-$ is a constant of contact hyperfine coupling determined by Fermi formula; $\langle S_z\rangle -$ is a characteristic of magnetic properties of the paramagnetic complex; g_n- is a nuclear $g-$factor; β_N – is a nuclear magneton. For a paramagnetic complex with insignificant spin-orbital coupling, the expression (1.4) looks as follows [5]:

$$\omega_k = -A\frac{\gamma_e g_e \beta S(S+1)}{3kT\gamma_N}, \tag{1.5}$$

where γ_e and γ_N – are gyromagnetic ratios of electron and nucleus; $g_e - g$ - is electron factor; S – is spin of the complex; β - is Bore magneton; T – is absolute temperature of the sample; k – is Boltzmann's constant.

The study of contact coupling in the second order of the perturbation theory [46] has revealed an agreement between the calculated values $\langle S_z\rangle$ and experimental ion-induced paramagnetic shifts in the ^{17}O and ^{14}N NMR spectra. The constant A turns out to be negative for ^{14}N and positive for ^{17}O, but it is independent upon nature of the lanthanide ion. The results obtained in [46, 47] were used for the evaluation of relative contact and pseudo-contact contribution to the LSR-induced paramagnetic shifts [48].It is established that among lanthanide diketones, europium ion should make the highest contact contribution, while the contribution of ytterbium ion is insignificant. The same conclusions have been drawn in [49, 50].

The detailed study of contact shifts in ^{13}C NMR spectra has been performed in [51]. Pyridine, 3,5-lutidine and isoquinoline were used as ligands. It is inferred [51] that $Eu(DPM)_3$

induces both contact and pseudo-contact shifts in the NMR spectra of pyridinic bases, the contact contribution in the ^{13}C NMR spectra being quite significant. At the same time, proton paramagnetic shifts can be successfully rationalized with a help of the pseudo-contact model. The value of contact contribution to paramagnetic shifts of signals from $Eu(DPM)_3$ and $Pr(DPM)_3$ in ^{13}C NMR spectra of exo-bornylamine has been assessed [52]. The authors share the common opinion that the shifts in ^{13}C NMR spectra are more sensitive to the contact coupling that those of protons. Therefore, the determination of values and signs ω_K for ^{13}C signals was preceded by calculation of the geometrical parameters, characterizing relative positions of lanthanide–ligand, from paramagnetic shifts in ^1H spectra assuming the existence of dipole-dipole hyperfine coupling (axial-symmetric approximation). Then the pseudo-contact contribution for ^{13}C was calculated. The difference between this contribution and the observed shift was considered as a contact component. The contact contributions from $Eu(DPM)_3$ and $Pr(DPM)_3$ appeared to be opposite in sign. It is believed [53] that there are two fundamental reasons explaining the fact that the contact contribution for protons should be significantly lower than for heavy nuclei.

1) The first one is a considerable coefficient of 2*s*-orbital located in the wave function of unpaired electron. This reason causes the high contact shift in signals of atoms (for example, nitrogen), with which the ligand molecules coordinate with lanthanides;
2) The second one is higher $[\psi(0)]^2$ for 2*s*-orbital as compared with 1*s*-orbital. Therefore, even when the coefficients defining the contribution of 2*s*- and 1*s*-orbitals to the wave function of unpaired electron are equal, the contact contribution to the paramagnetic shifts should be considerably higher than that to the proton signal shifts.

However the conclusions drawn in the work [53] allow explaining the different contribution of the contact shifts for hydrogen nuclei (protons) in comparison with nuclei of other atoms. If to take into account the results of the investigation of hyperfine coupling on ^{13}C and ^1H nuclei in paramagnetic complexes of iron group elements, the above reasons do not explain why proton paramagnetic shifts in the PMR spectra of paramagnetic complexes of lanthanide diketons should be less sensitive toward the contact coupling interaction than those in the ^{13}C spectra.

In the discussed works, the studied ligands belong to the systems where contact HFC is expected to be transferred across the $\sigma-$system. It is assumed [54] that in the ^1H, ^{19}F and ^{13}C NMR spectra of the substrates with the expected delocalization at $\pi-$system, LSR also induces considerable contact shifts. Such inference is made from the discussion of experimental values of paramagnetic shifts using the ideas on delocalization mechanism of uncompensated spin density in paramagnetic complexes of iron group elements. The contact contribution depends upon both ion nature and chelate and increases as follows:

$$Pr(FOD)_3 < Pr(DPM)_3 < Yb(DPM)_3 < Yb(FOD)_3 < Er(DPM)_3 < \\ < Eu(DPM)_3 < Er(FOD)_3 < Eu(FOD)_3 < Eu(PTA)_3 .$$

It is attempted to quantitatively evaluate the π -contact component of the paramagnetic shifts induced by $Eu(FOD)_3$and $Pr(FOD)_3$in ^{1}Hand ^{13}C NMR spectra of analinine and p - toluidine [55]. The relative values of contact shifts calculated by MacLachlan method were used for this purpose. Minimum value of agreement factor R was considered as a criterion of calculation accuracy [56]:

$$R = \left[\frac{\sum(\omega_{obsr} - \omega_{cacld})^2}{\sum(\omega_{obsr})^2} \right]^{1/2}.$$

The values $\omega_{cacl(}$ were calculated by McConnell–Robertson formula in axial - symmetric approximation [57]. The analysis of experimental data has shown that in $Eu(FOD)_3$and $Pr(FOD)_3$complexes, the transfer of negative and positive spin density from the coordination ion to the ligand can take place. But if to bear in mind the exact quantitative calculations, this conclusion should be checked, since it is stated [55] that it is not clear whether the procedure used for determination of R accounts for the contact contribution to the calculated values of paramagnetic shifts. The problems of separation of contact and pseudo-contact contributions to LSR-induced paramagnetic shifts are also discussed in [58-60].

If the reality of the contact contribution to the LSR-induced shifts (at least, in other than ^{1}H NMR spectra) is accepted by majority of researchers, the mechanism of contact coupling transfer in paramagnetic complexes of lanthanide diketones with organic ligands still remains controversial. It is found [61] that the $\pi -$coupling mechanism cannot lead to considerable contact shifts. Abnormal shifts for ^{1}H signals are likely due to the direct through-space induction of spin density on resonance nuclei, or follow from the inaccuracies in calculation of pseudo-contact shifts. However, it is argued [50] that the Eu···N bonding in the $Eu(FOD)_3$ complexes with pyridine and its derivatives is definitely of covalent nature. This leads to the appearance of positive spin density on the ring protons, and, hence, the contact contribution to the paramagnetic shifts. Notably, the substituent in the position 2 increases the uncompensated spin density on H-3 relative to H-5 like in tetrahedral complex CoX_2L_2 (L - 2-methylpyridine). On the example of quinoline, it is shown [59] that lanthanide trisdiketonates induce contact shifts in the ^{1}H and ^{13}C spectra, their values being dependent on Ln ··· N covalence degree.

According to the quantum-chemical calculations (CNDO/2) [62], the contact shifts in the ^{13}C NMR spectra of merocyanines are due to the spin density transfer across the system of π -bonds. It is noted [55] that in the europium complex there are two possibilities of spin density delocalization on ligands: 1) the $4f$ -orbital, which is not occupied by the unpaired electron, is overlapped with the orbital of nitrogen atom lone pair. In the result, electron spins of this pair interact (through spin exchange) with uncompensated electron spins of other f - orbital; 2) the orbitals of ligand and coordination ion are not overlapped, and the contact shift results from electron spin polarization of lone pair of the coordinated ligand atom with uncompensated electron spin of the europium ion. Both mechanisms should lead to localization of insignificant amount of unpaired electron spin on nitrogen atom and this is supported by experimental data. To explain the observed values of paramagnetic shifts, one should assume that ~ 10^{-3} of the ligand negative spin density is located in π -system of the ligand. In the praseodymium complex, the overlapping of $4f$ -orbitals, occupied with unpaired electrons, and orbitals of nitrogen lone pair is possible. However, in this case, one should admit that in π -system of the aniline cycle the localization of unpaired density is not higher than in the europium complex. Therefore, it is unjustified to assume the overlapping of $4f$ -orbitals of the central ion, occupied with unpaired electrons, and the ligand orbitals. Earlier, it has also been concluded [46] that $4f$ -orbitals do not participate in direct coupling with the ligand orbitals.

The nature of contact shifts in the ^{13}C NMR spectra of pyridine and β -pycoline induced by $Eu(DPM)_3$,$Pr(DPM)_3$,$Eu(FOD)_3$, has been discussed [63]. For this purpose, the experimental shifts of ^1H and ^{13}C signals are compared with relative pseudo-contact shifts. While the experimental values of paramagnetic shifts for ^1H are explained by the pseudo-contact mechanism, for ^{13}C nuclei full inconsistence between calculated and experimental values is observed. It is inferred that if even π -contribution takes place, its value is very small and other type of delocalization mechanism, should be realized. The contact shifts, induced according to this mechanism, should be significantly higher for ^{13}C as compared to ^1H.

Comparison of paramagnetic shifts in the ^1H and ^{13}C NMR spectra of pyridines induced by $Ni(AA)_2$иEu$(FOD)_3$, shows that ω_K values for the corresponding nuclei are comparable, but their signs are opposite, i.e. transfer of the contact coupling across the ligands in the $Ni^{2+}$and Eu^{3+}complexes proceeds in a similar fashion (σ -delocalization) [64]. Distinctions in signs are owing to different mechanisms of spin density localization on nitrogen atom. In the $Ni(AA)_2$ complex, the unpaired electron is delocalized from d_{z^2} -orbital of the central ion directly on nitrogen atom that leads to appearance of positive spin density. In the $Eu(FOD)_3$ complex, unpaired $4f$ -orbitals induce positive spin density on 6*s*-orbital, thus causing the negative spin density on nitrogen atom (spin polarization mechanism).

The nature of Fermi interaction transfer on the ligands of pyridinic type has also been discussed [51]. It is concluded that the contact shifts induced by $Eu(DPM)_3$ can be rationalized using the inferences made in [18] concerning the mechanism in σ -systems. It is assumed that the complex formation with europium ion leads to β -uncompensated electron spin in σ -system of the ligand (instead of $\alpha -$ spin as it is observed in nickel complexes). This supports satisfactory agreement between theoretical and experimental values of relative contact shifts.

Thus, if to postulate the localization of uncompensated electron spin density (having the corresponding sign) on ligands, one can explain the experimental data concerning signs and relative values of the LSR-induced contact shifts in the NMR spectra of compounds, in which one should expect the realization of σ -delocalization mechanism. However, even in these cases, the mechanism of hyperfine coupling transfer from lanthanide ion to ligands still remains debatable.

1.2. Study of Spatial Structure of Organic Compounds

Characteristic dependence of the paramagnetic shifts on relative position of the coordination ion and resonance nucleus makes it possibly to apply the coordination ability of heteroatom compounds for studying their spatial structure. However, up to 1970, the investigations in this direction were sparse. The overwhelming number of papers on application of the paramagnetic complexes has appeared after publications of Hinckley [7, 65], Sanders and Williams [8]. It is reported [7] on comparability of the PMR spectra of cholesterol solution recorded without and with additives of the paramagnetic $Eu(DPM)_3{\cdot}2Py$ complex. The use of the latter is explained by very short time of electron relaxation (see Chapter 2) of rare-earth element ions. Therefore, the complex formation leads to insignificant broadening of resonance lines, whereas times of their paramagnetic shifts are considerable.

The ability of $Eu(DPM)_3{\cdot}2Py$ to induce relatively high paramagnetic shifts, which depend upon concentration of the complex in the studied sample without essential broadening of resonance lines, allows to conclude that this complex can be used as a shifting reagent. It is reported [65] on the application of europium paramagnetic complex for the assignment of methyl proton signals in the PMR spectrum of ***d*** -camphor though in literature the opinions differ widely on this issue. The spectral analysis is premised by the following rationalization of paramagnetic additives usage. Upon dissolution of $M(DPM)_3{\cdot}2Py$ complex (M is a rare-earth element) in CCl_4, the its molecule dissociates according to the equations:

$$M(DPM)_3{\cdot}2Py \rightleftarrows M(DPM)_3{\cdot}Py + Py,$$

$$M(DPM)_3{.}Py \rightleftarrows M(DPM)_3 + Py.$$

If organic compound B having the coordination group, is added to this solution, the following associations occur:

$$M(DPM)_3{\cdot}Py + B \rightleftarrows M(DPM)_3Py{\cdot}B.$$

Thus, paramagnetic shifts observed in the spectrum of compound B result from such association.

As stated in [65], the application of the rare-earth element complexes should include the following procedures: 1) recording the spectrum of compound B dissolved in CCl_4 with one or several $M(DPM)_3{\cdot}2Py$ additives; determination of concentration dependence of the observed shifts is required when spectral analysis is difficult at one concentration; 2) modeling of compound B and measuring the distances from coordinating metal ion to resonance protons; interpretation of compound B spectra.

It is shown [8] that $Eu(DPM)_3$ is also an excellent shifting reagent. Other complexes of rare-earth elements turn out to be useful for this purpose [66-73]. The problem of rational selection of LSR for various substrates has been discussed [66]. The investigations aimed at the synthesis of novel shifting reagents are conducted. It is disclosed that macrocyclic systems can be applicable for these purposes [74,75]. In this line, tetrakis(1,1,1-trifluoro-4-phenylbutane-2,4dianato)uranium is found to be efficient, since in the spectra it causes significant shifts with conservation of their multiplicity. Besides, it is well soluble and nonhygroscopic [76]. It is proposed [77] to apply $AsCl_3$ as a shifting reagent (together with LSR) and also as a solvent upon usage of paramagnetic additives. It is assumed [78] that equimolar mixture of LSR and $C_3F_7CO_2Ag$ can be useful to for the NMR analysis of alkenes, i.e. the compounds which are not coordinated with paramagnetic shifting reagents. The details of such additives application are given in [79, 80]. A special study [81] deals with binuclear shifting reagent $AgLn(FOD)_4$, and its interaction with benzene. It is shown that the complex formation proceeds via the interaction with Ag^+ followed by the formation of $Ag^+(FOD)_4^-$ ion pair. As shown in the scheme below, $Yb(FOD)_4^-$ is located over the aromatic ring [81]:

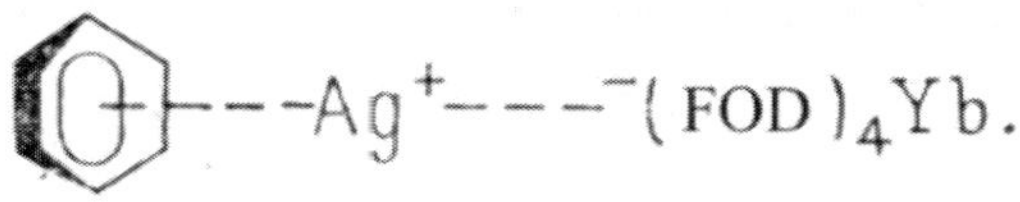

It is suggested [82,83] to use $TiCl_4$ as a shifting reagent for assignments of signals in the NMR spectra of organic compounds containing polar functional groups. It is concluded that, depending on ability to be coordinated with $TiCl_4$, the studied compounds are ranged in the following order: alcohols, ethers, esters It is also found that the adduct of $TiCl_4$ to substrate formed in the solution has 1:2 composition. Applicability of $TiCl_4$ for ^{13}C NMR spectral analysis is shown on an example of β -lactams containing the carbonyl group in the four-membered cycle. Similarity of signs and values of the induced shifts allows one to deduce that $TiCl_4$ can be used as a reagent for the detection of carbonyl carbon signals. Advantages and shortcomings of $TiCl_4$ application as a shifting reagent have been discussed [83].

If to mean only simplification of the NMR spectral analysis, the expansion of shifting reagents scope, of course, increases possibilities of the researchers. However, successful use of these reagents for the solution of structural tasks is defined by understanding of the nature of the shifts induced by these reagents.

1.2.1. Complexes of First Transition Group Elements

According to the NMR theory, in paramagnetic systems, contact and pseudo-contact* shifts ω'_p are proportional to $1/T$, i.e. obey to the Curie law. Therefore, the expressions $\omega''_k = \omega_k \cdot T$ and $\omega'_p = \omega_p \cdot T$ should not depend upon temperature of the studied sample. Existence of such temperature dependence may evidence either the alteration of spatial location of the resonance nucleus relative to the coordination ion or the change of uncompensated electron spin contribution on the nucleus. This circumstance has been used for investigation of conformational equilibrium in nickel chelate complexes coordinated with 1,3-diamines [84]. Taking into account the steric effects, among three conformations of the above complexes, "chair" and "twist" ones should be realized. Provided that the time of Ni···N bond existence is much higher (in the NMR scale) than the time characterizing the chelate

*We mean here pseudo-contact interaction described by McConnell-Robertson formula [57].

ring inversion (owing to conformational transitions), the observed $\omega \cdot T$ values are expressed as follows:

$$\omega \cdot T = \sum x_i (\omega \cdot T)_i, \tag{1.6}$$

where x_i – is a molar share of i – conformation which is characterized by $(\omega \cdot T)_i$ value.

If to make a reasonable assumption concerning ω or to use paramagnetic shifts for the similar compounds, the experimental $\omega \cdot T$ values make it possible to determine a contribution of the corresponding conformation, while character of temperature dependence of the paramagnetic shifts allows finding the thermodynamic parameters of the conformational equilibrium. Applicability of expression (1.6) is illustrated on the example of Ni (2,4-diaminopentane) [84].

It has been emphasized that the isotropic constant of hyperfine coupling of β - protons with unpaired electrons in π -electron radical is defined by the expression (1.7) [85]:

$$A_\beta = Q(\theta) \cdot p, \tag{1.7}$$

$$Q(\theta) = C' + D\cos^2\theta, \tag{1.8}$$

where p – is a spin density on p_z – orbital of the neighboring carbon atom; C' and D – are constants; θ – is an angle formed by projections of C – H_β bond and p_z – orbital axes on the plane perpendicular to the C-C bond direction as shown in Scheme below.

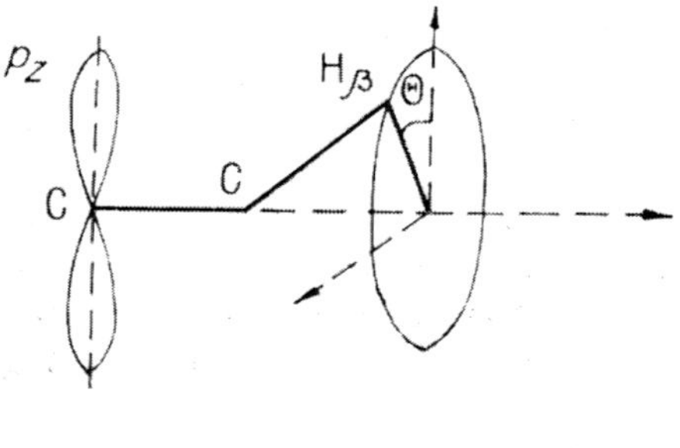

If to use the formula for contact shift and to assume that the mechanism of contact coupling transfer in the above paramagnetic complex is similar to that in the radical fragment, which is described by the equalities (1.7) and (1.8), then

$$\omega_k = C' + D\cos^2\theta. \tag{1.9}$$

Usually, $C' << D$, therefore, it is supposed [84] that $D\cos^2\theta >> C'$. The parameters C' and D have been analyzed [85-88]. Applicability of the expression (1.9) for conformational analysis using the hyperfine coupling constants of ^{13}C nuclei is discussed [89].

The PMR spectra of pseudo-tetrahedral complexes V (see scheme below) have been recorded at different temperatures [90], the equality (1.9) also being successfully used the conformational analysis. It is state that the following expression is the most appropriate:

$$\omega'_{\kappa} = \omega_{\kappa} \cdot T/298 = -60400 \cdot \rho . \qquad (1.10)$$

The value ω'_{κ} (ppm), normalized relative to temperature of 298 K, depends only upon the spin density of unpaired electron on the proton.

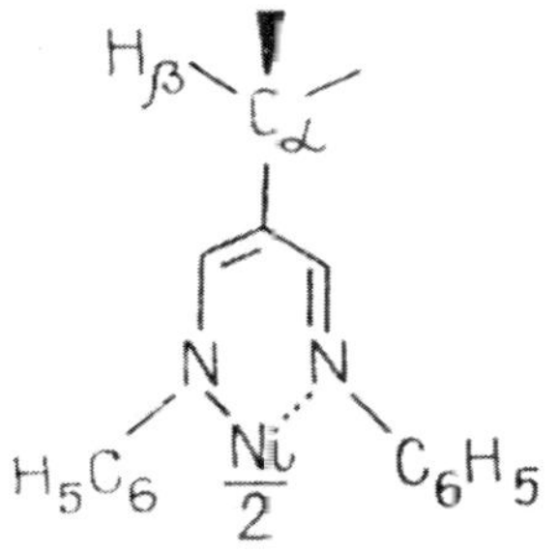

V

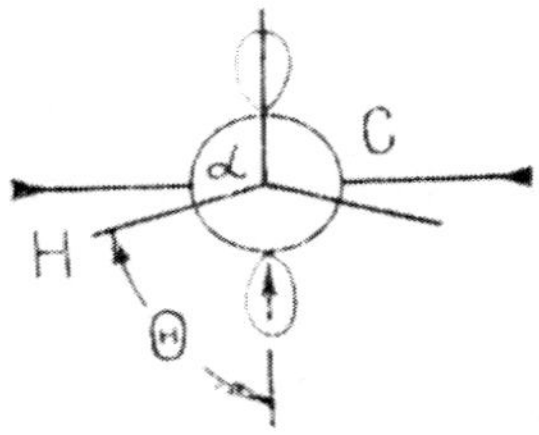

VI

For five shifts from six, the ω'_{κ} value does not depend upon T. Only ω'_{κ} value for the signal of two equivalent β -protons increases (in absolute magnitude) with growth the studied sample temperature. This fact is explained by the alteration of ρ value on β -nuclei due to rotation around the C – C bond (structure VI). The experimental data are processed taking into account that in expression (1.9), the value of cosine should be averaged relative to all possible rotations of θ angle. The results obtained have been compared with the corresponding parameters found by other methods (ESR-spectroscopy, theoretical calculations, and microwave spectra) for the compounds modeling hyperfine coupling transfer across the ligands of complexes V.

The examples given above confirm the common fact of high sensitivity of the contact hyperfine coupling towards stereochemistry of ligand molecules [31, 91]. However this dependence still has not found broad application in spite of the fact that it is useful, for example, for the study of such fine effects as $n-\pi$ -transannular interaction [92]. The difficulties arising owing to considerable broadening of resonance are among the objective reasons hampering the application of contact paramagnetic shifts. Nevertheless, one should admit that in the presence of such huge amounts of synthetic and natural chemical compounds, the application of unpaired spin density delocalization on and across the ligands is unjustifiably limited. We believe that the progress in the NMR spectrometers will also expand a circle of tasks which solution is connected with the specifics of NMR phenomenon in paramagnetic systems. The modern NMR spectrometers operating in a pulse mode allow recording the spectra of paramagnetic systems in wide spectral ranges (in particular, about 400 ppm for ^{1}H NMR and 2000 ppm for ^{13}C NMR). High-speed workstations of such spectrometers are capable of processing huge massifs of 2M spectra that makes it possible (at least, in principle) to assign the resonance signals. The application of temperature devises allows carrying out complex dynamic researches involving paramagnetic systems.

The intriguing investigations involving the NMR phenomenon in paramagnetic systems have been carried out by the Rostov researchers under the supervision of academician V.I. Minkin. They study, in particular, stereodynamics of tetracoordinated chelates of transition metal complexes and ligand exchange processes [93]. A series of works is devoted to the HFC-transformed ^{13}C NMR spectra of outer-sphere complexes [94]. It has been shown that such complexes can play an important role in diverse catalytic processes. It has been found for the first time [95] that the broadenings of signals, caused by paramagnetic shifting reagents, can be successfully used in structural researches (see Chapter 2). In this line, we can refer to work [96], where the conformations of various amino acids coordinated with paramagnetic ions Mn^{2+}, Cu^{2+}, Co^{2+}, Ni^{2+}, Fe^{3+}, Cr^{3+}have been investigated.

It is offered to use $Cu(AA)_2$ as a broadening selective reagent for the assignment of ^{13}C signals [97]. In other paper [98], instead of signal broadenings, the relaxation times of resonance nuclei, obtained by direct measurement, have been employed in structural studies. It is noted that the paramagnetic relaxation reagents (PRR) can be used for (i) facilitation of non hydrogen-coupled ^{13}C signals detection; (ii) elimination of nuclear Overhauser effect in $^{13}C - {}^{1}H$ in quantitative experiments involving small molecules; (iii) suppression of nuclear Overhauser effect in experiments dealing with the study of spin-spin coupling of protons and nuclei having negative gyromagnetic ratio. The relative disposition of resonance nuclei of ligand and coordination ion can be determined by the expression describing the measured time T_1' of spin-lattice relaxation:

$$\mathbf{1/T_1' = 1/T_{1M} + 1/T_1}, \tag{1.11}$$

where $\boldsymbol{T_1}$ – is controlled by diamagnetic mechanisms of a spin-lattice relaxation in the sample containing no PRR, and $\boldsymbol{T_{1M}}$– is controlled by hyperfine coupling. The first member of the expression (1.11) has two contributions, i.e.

$$\mathbf{1/T_{2M} = P_M / T_{1M}^k + P_C / T_{1M}^c}, \tag{1.12}$$

where $\boldsymbol{T_{1M}^k}$ and $\boldsymbol{T_{1M}^c}$ relate to the ligand nuclei in and out of first coordination sphere of the complex (when PRR are present in the sample); $\boldsymbol{P_M}$ and $\boldsymbol{P_c}$ – are molar fraction of the corresponding states. $\boldsymbol{T_{1M}^c}$ value is identical to all nuclei of the molecule and can be determined from the NMR spectra of the compound (structurally similar to the ligand), the molecule of which is out of first coordination sphere of the paramagnetic ion.

To obtain structural and dynamic information, the authors of the work [99] have measured relaxation times $\boldsymbol{T_1}$ and $\boldsymbol{T_2}$ in the presence of $MnCl_2$. It is revealed that spin-lattice relaxation is controlled by dipole-dipole coupling, while spin-spin relaxation is defined by scalar (contact) coupling. $\boldsymbol{A}$ Constants between unpaired electrons and ^{13}C nuclei, activation parameters of the ligand molecules exchange, and distances between coordination ion and the ligand carbon atoms (histidine) have also been experimentally determined.

The examples discussed in this section allow concluding that the application of paramagnetic complexes of iron group elements in structural studies can be quite effective.

1.2.2. Complexes of Rare-Earth Elements

By the moment, about two thousand papers dedicated to application of the rare-earth element complexes for the analysis of NMR spectra have been published. All these papers can be divided into two groups.

The first group includes the works, where LSR play a "passive" role, i.e. they are employed only for simplification of the NMR spectra shape. In this line, the application of LSR together with partial deuteration, alteration of the studied sample temperature and resonance frequency of the NMR spectrometer, as well as the usage of double resonance technique and Overhauser intramolecular effect turn out to be advantageous. The spectra thus simplified are usually analyzed according to the rules for the spectra of first order. The found values of chemical shifts and ***J*** constants are further used for the study of substrate molecules. This group also contains the papers devoted to determination of enanthiomeric structure. The application of LSR appears to be fruitful for NMR studies of the dynamic phenomena. Indeed, LSR can augment the difference of chemical shifts of the resonance nuclei in various chemical media, i.e. change the NMR time scale relative to the studied process. Finally, numerous works cover the following issues: 1) the complex formation of LSR with various classes of compounds; 2) efficiency of this process and steric effects; 3) additivity of paramagnetic shifts for compounds having several coordination centers, etc. The efficacy of the LSR usage in the "passive" purposes for structural studies of complex substrates (such as, for example, diamagnetic complexes VII) has been shown in the review [100].

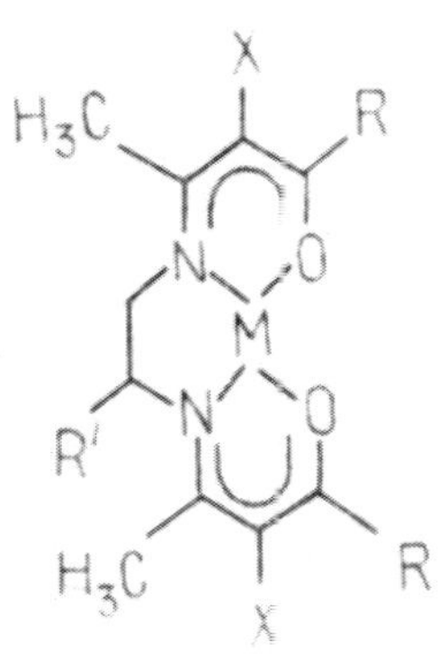

M = Ni, Pd, Pt
R = alkyl
R′ = H or CH_3
X = H, Cl, Br, I,
CH_3, SC_6H_5.

VII

The second group includes the works, in which lanthanide shifting reagents play an "active role", i.e. the value of shifts and signal broadenings (or relaxation times) are considered as a source of information on spatial structure of molecules coordinated with paramagnetic ion. To the above stated concerning the application of rare-earth element complexes, one should add that they are effective, when are used both in "active" and "passive" purposes, for studying the biological systems. The application of LSR is unexpectedly fruitful for the analysis of the NMR spectra of oils. The difficulties connected with investigations (including by the NMR method) of these compounds are obvious. ^{13}C NMR spectra of light oil fractions (gasoline and diesel), for example, contain several dozens of signals, which identification represents a problem. We do not speak about the use of such spectral information in structural studies. Nevertheless, it has been shown that both individual LSR [101] and binuclear paramagnetic complexes [102] can be successfully used for study of, at least, some oil fractions.

In this connection, the structural studies involve the determination of the pseudo-contact shift values by the following expression [1,2]:

$$\omega_p = D_1 \frac{3\cos^2\theta - 1}{r^3} + D_2 \frac{\sin^2\theta\cos\phi}{r^3}, \qquad (1.13)$$

where D_1 и D_2 – are functions of magnetic anisotropy of the paramagnetic ion; θ – is the angle between z axis and the direction to the nucleus, for which the NMR spectrum is observed; r – is the distance from this nucleus to the coordination ion; ϕ – is the angle between vector r projection to the xy plane and axis y. For complexes with axial symmetry $D_2 = 0$, therefore, the expression (1.13) is replaced by McConnell-Robertson formula:

$$\omega_p = D_1 \frac{3\cos^2\theta - 1}{r^3}. \qquad (1.14)$$

Though it has been assumed [7, 103] that it is not evident that the second member of the expression (1.13) can be neglected in the comparison of the theoretical and experimental ω_p values induced by LSR, the overwhelming number of researches uses the expression (1.14). Besides, in the first works related to the application of LSR in structural studieshes, the formation of LS complexes (L – a lanthanide, S – substrate) are reported. Consequently, the coincidence of the main magnetic axis and lanthanide –substrate bond is considered as natural. The researchers, using the model of axial-symmetric complex in solution, discuss a need of the search for such position of the magnetic axis, which would provide the best agreement between the calculated and experimental shift values. Results of the magnetic susceptibility measurements and X-ray diffraction analysis of lanthanides tris(diketone) adducts with organic ligands throw discredit upon the possibility of the expression (1.14) usage in structural studies [104]. However, in solution owing to the rotation of substrate molecules around the axis passing through a lanthanide, there will be such averaging of the shifts that the axial symmetry model becomes applicable, and the θ angle in the expression (1.14) should be calculated taking into account this effective symmetry axis. To average the second member of the expression (1.13) to zero, at least three rotary isomers with identical population should be formed [105, 106].

The numerous publications, reporting specific results of the structural studies using the model of axis-symmetric complex, can be considered as an illustration of successful application of the expression (1.14). Nevertheless there are no grounds to claim that such approach is self-consistent. For example, direction of the Ln···O bonding, obtained by processing of the induced shift values, considerably and chaotically deviates from the bisecting line of the C – O – C angle in ethers and C – O – H angle in alcohols as well as from direction of the C = O bond in carbonyl compounds. Additionally, the Eu···O distance ranges 2,1 – 4,0 Å [107]. Arbitrary spread of Eu ···N distances is observed at interpretation of the LSR-induced paramagnetic shifts in the spectra of nitrogen-containing compounds [107]. All this contradicts the fact that ability of organic base to interact with LSR is caused by lone electron pair (oriented in well-defined manner) of the substrate donating atom.

Horrocks, having paid attention to the hindrance of rotation around lanthanide-substrate bond due to steric factors, has offered a model rationalizing the application of the expression (1.14), despite the lack of axial symmetry in crystalline complexes [108]. According to the known stereochemical mobility of seven- and eight-coordinated compounds, it is supposed that the LSR-substrate adduct in solution represents an ensemble of a large number (about 30 and more) of geometrical isomers, between which fast (in the NMR scale) transitions take place. Each of the isomers is characterized by its own magnetic susceptibility tensor, and the substrate molecule is arbitrarily oriented relative to main magnetic axis, but in the ranges specified by additional conditions. For two typical substrates, cyclohexanone and isoquinoline, this model gives the values of the relative shifts, which are close to the experimental. It is noted [108] that ω_p values corresponding to the observed pseudo-contact shifts can be obtained with a help of the axial symmetry model by varying θ and r in the expression (1.14). However in this case, the lanthanide-substrate distance corresponding to R_{min} can differ by 0,5 Å and is higher than the distance which is really realized in the complex.

The issue on accuracy and applicability of the axial symmetry model for signal broadenings has been discussed [109]. In this connection, the of values of signal broadenings caused by $Ln(DPM)_3$ (Ln=Er, Ho, Dy, Tm, Gd) DPM) in the spectra of 3,5-lutidine are compared. It is found that the assumption about axial symmetry, made upon interpretation of signal broadening values, is not quite correct. The authors of the discussed work suppose that the latter remark should not be considered as a prediction of inefficiency of the axial symmetry model in structural researches using pseudo-contact shifts.

If the magnetic anisotropy tensor is not axially symmetric, the second member in the expression (1.13) can be averaged to zero due to the exchange or rotation. However, the corresponding member for relaxation time in this expression is not averaged to zero. Therefore, the signals broadening will contain an additional contribution owing to failure of the axial symmetry conditions, when the shifts will not have such a component. This contribution is confirmed by the fact that the different averaging rate of the observed shifts and broadenings is inherent in these parameters (see Chapter 2). Therefore, the measurement of broadenings at different temperatures seems to be necessary, though it is considered [109 that independence of signal broadening values in *ortho*- and *para*-protons upon the LSR/substrate molar ratio testifies to insignificance of the chemical exchange contribution to the resonance lines width.

The dependence of ω_p upon angular factor has prompted the authors of the works [110, 111] to apply the paramagnetic relaxation reagent (PRR) having an isotropic g – tensor for analysis of the NMR spectra. It is disclosed [111] that relaxation times of the ^{13}C nuclei, measured in presence of paramagnetic relaxation reagent $Gd(DPM)_3$, are used for assignment of the signals in the ^{13}C NMR spectra. The applicability of the method proposed has been shown on the example of γ-picoline. The contributions from other relaxation mechanisms (not connected with the effect of paramagnetic relaxation reagent) are excluded from the comparable values. Decrease of the relaxation time T_1 (owing to outer-sphere effects) is also taken into account. For this purpose, the relaxation times of carbon nuclei of toluene, which does not coordinate with $Gd(DPM)_3$, are measured. It is stated [111] that one can avoid the considerable spread in values if to apply non-coordinating relaxation reagent $Cr(AA)_3$ (see, however, the section 1.2.1 where the work [98] is discussed). The applicability of gadolinium chelate for interpretation of the PMR spectra using the signal broadenings is considered [110]. The same work reports the combined application of LSR and PRR; the shifting reagent, for example, $Eu(DPM)_3$, in this case increases the difference of the chemical shifts, which are further assigned by the PRR technique.

Here it is pertinent to make the following comment. The application of the standard every time can "hide" the peculiarities of the relaxation processes inherent in a specific compound. Besides, if it is not a problem to select such a standard for γ- picoline, for more complex compounds, this procedure may meet with difficulties. Nevertheless, the combined application of LSR and PRR, as it is noted in [110], should improve the accuracy of the results of structural studies using the paramagnetic complexes. When PRR is employed, one should bear in mind that the broadening of signals with multiplet structure can occur due to the removal (suppression) of spin- spin coupling by chemical exchange. In this case, one should apply the technique of double resonance to avoid the broadening owing to the above effect [112].

In the context of the above-discussed models, the study of luminescence spectra of $Eu(DPM)_3$ and its some adducts [113] are of special interest. In particular, it has been established that:

1) The configuration of crystalline eight-coordinated $Eu(DPM)_3 \cdot 2Py$ complex (Ru – pyridine) remains intact upon at dissolution;
2) The adduct of borneaol with $Eu(DPM)_3$ (1:1 composition) is configurationally labile and has no exclusive predominant geometry in the solution. Therefore, this complex is characterized by fast transitions between structures VIII and IX shown below.

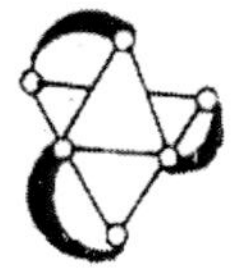

VIII

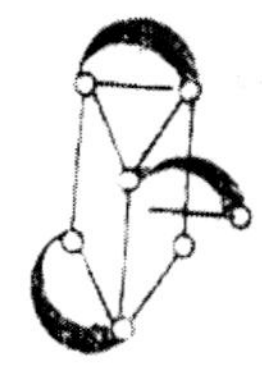

I X

Results of the work [113] allow to assume that for the LSR adducts the comparison of theoretical and experimental shifts should provide application of the expression (1.13). In the spectra of the complexes, when the substrates have no symmetry axis of the third order, the intraconfiguration transitions could be realized. These transitions ensure the averaging to zero of second component of the expression (1.13) due to either the rotation around the lanthanide-ligand bond or realization of the Horroks model.

Unlike the majority of researchers, the authors of the work [114] believe that the molecular structure of diadduct $Eu(DPM)_3$ preserves in the solution. Starting from this assumption, they have analyzed the PMR spectra of pyridine substrates recorded with additives of $Eu(DPM)_3$. Low-temperature measurements indicate the formation of 1:2 adducts, which remain intact at higher temperatures. The values of experimental shifts are compared with the calculated (according to the expression (1.13)) shifts using the geometrical parameters of crystalline complex $Eu(DPM)_3 \cdot 2Py$. In the calculations, protons of both substrate and DMP are considered. Diamagnetic corrections ω_g are subtracted from the induced shifts. As a consequence, the good agreement between the theory and experiment has been achieved. In the result, it has been emphasized [114] that the expression (1.13) should be applied for quantitative estimations of ω_p induced by LSR.

The application of the expressions (1.13) and (1.14) in structural studies should involve the determination of ω_p value relating to a complex of a certain structure. Usually the experiments are carried out with excess (relative to LSR) content of the substrate in sample. General aspects of the steriochemial issue have been discussed in the work [115] on the basis of the two-step mechanism of the complex formation:

$$L + S \xrightarrow{K_1} LS, \tag{1.15}$$

$$LS + S \xrightarrow{K_2} LS_2, \tag{1.16}$$

$$\varepsilon = \frac{1}{[S_0]}([LS]\omega_1 + [LS_2]\omega_2), \tag{1.17}$$

where K_1 and K_2 are the association constants for processes (1.15) and (1.16). ω_1, ω_2, K_1 and K_2 values can be calculated from the values of experimental shifts for various concentration of $[S_0]$ and $[L_0]$ using the minimization of criterion Q [115]:

$$Q = \sum_{i=1}^{n} (\varepsilon_i - \delta_i)^2 \tag{1.18}$$

,

where δ_i and ε_i – are experimental and calculated (equation 1.17) shifts, respectively; n – is a number of observations differing in $[S_0]$ concentration and (or) $[L_0]$.

It has been noted [115] that the application of LSR in structural studies should account for the values of shifts obtained by statistical analysis of the dependence of δ upon $[L]/S$ including $[L]/[S] > 1$, and not for the data related to linear range of the question

$\delta = f([L]/S)$. This conclusion is based on the fact that the solution should contain both adducts, which are characterized by different K and ω values. However, there is a popular opinion that when substrate is used in the excess concentration and under the conditions of fast exchange, the main contribution to the induced shifts is made by the adduct only. In this regard, the paper [116] is of particular interest. The authors, using the effect of induced circular dichroism, have shown that under standard conditions of the NMR spectra recording ($[L]/[S] > 2$), the compound Eu(FOD)$_3$ forms a complex of 1:2 structure.

As for the discussion concerning structure of the complex responsible for induction of paramagnetic shifts in the NMR spectra used in structural studies, it is pertinent to make the following remark. As it was already noted, the observed paramagnetic shifts are considered as weighted average from two values characterizing free and coordinated molecules. If determination of the chemical shift characterizing the first (non-coordinated) molecule meets with no problems, the estimation of signal positions in the NMR spectra of resonating nuclei incorporated in the coordinate molecule faces certain difficulties. Apparently, in this case, the problems become more serious due to formal application of the exchange processes theory to the method of paramagnetic additives. Indeed, the starting point is postulation of the fact that the solution should contain, at least, two complexes. But, after all, the complex formation process imposes certain requirements on the ligand molecules, so that the latter (according to the symmetry rules and depending upon type and energy of the central ion orbitals) could form the coordination sphere. And if for the specific state of the ion (for example, M^{+3}), the corresponding surrounding is favorable, liabilty of the complex does not mean its destruction. In other words, owing to indistinguishability of the molecules, they randomly (chaotically) enter and leave the first coordination sphere of the paramagnetic ion bearing information on HFC, and it is reflected in the average spectrum.

It is a common knowledge that metal M (rare-earth element) in M(DPM)$_3$ or M(FOD)$_3$ increases the coordination number to eight. Therefore, it is natural to consider that under the corresponding conditions, the complex containing two molecules of the substrate is formed in the solution. That is why the Eu(DPM)$_3$·2Py complex has been employed in the work [7]. Certainly, d- or f-elements, which define specifics of adducts, can form different (in magnetic properties, stoichiometry, etc.) complexes with other compounds. Therefore, before application of these complexes for the analysis of the NMR spectra, it is necessary to elucidate their specifics. Another dramatic problem is in the fact that the molecules of each substrate should not mandatory form paramagnetic complexes with the corresponding symmetry of the central ion surrounding. However, the probability of complex formation of the weak bases can be increased via preliminary formation of the coordination sphere. In this case, if the coordination site in this sphere will be free for some time, the molecule of weak basic compounds most likely occupy it.

The PMR spectra of 1-methylsilatrane (S) recorded with additives of Eu(DPM)$_3$ (L) at different temperatures have been analyzed [117]. It is known that high basicity of the substrate impedes the realization of the process (1.15). On the contrary, low basicity, which is typical for weak Lewis bases, should shift the equilibriums (1.15) and (1.16) to the left, thus facilitating the process (1.15).

Table 1.1. Paramagnetic shifts of – CH_2 – и CH_3 – proton signals for different values of P ($t = 25\ ^0C$)

P	Absolute shifts, Hz				Relative shifts		
	Si – CH_3	CH_2 - O	CH_2 - O	t – Bu	Si – CH_3	CH_2 - O	CH_2 - O
0,2	0	0	0	0	0	0	0
0,5	7	2	2	- 29	1,00	0,30	0,30
0,8	67	25	17	- 75	1,00	0,37	0,25
1,0	106	40	27	- 79	1,00	0,37	0,25
1,3	129	48	32	- 79	1,00	0,37	0,25
1,7	183	68	47	- 78	1,00	0,37	0,25

Note: Minus corresponds to the high-field shift of the signal.

Table 1.1 shows the observed values of δ shifts at various $P = $L/S ratios in the sample studied [117]. Also, the normalized (relative to δ_{CH_3}) values of paramagnetic shifts are given in the Table. The transformations in the PMR spectra of substrate S are observed upon increasing the P value, which is defined by the comparison of integrated intensities of L and S signals. These transformations can be explained, if to admit that δ value of signals S is caused by the LS_2 complex, while δ of signal L ($t-$Bu) is due to the LS complex or both, but the shift in LS_2 is lesser in value and opposite in sign as compared to the LS complex. Therefore, δ values of the CH_3 and CH_2 groups of 1-methylsilatrane increase symbatically, while the value of the normalized shifts remains unchangeable. At the same time, δ for $t-$Bu, which reaches the highest value when $t-$Bu = 1, further does not alter. Since the shift of $t-$Bu signal becomes significant when $P >$ 0,2 and that of the S signals at even higher concentration of L in the sample, the processes given in the expressions (1.15) and (1.16) should be characterized by small K values, K_1 being higher than K_2. It has already been mentioned that paramagnetic shifts of the substrate nuclei are caused by the formation of LS_2 adduct owing to higher value of K_2 as compared to K_1, but not to the absence of shifts in the 1:1 complex.

The experimental data reported in the work [117] allow us to draw the following conclusions: 1) to induce paramagnetic shifts and broadenings in the NMR spectra of the substrate, the latter should form a complex of 1:2 structure with lanthanide trisdiketonates; 2) the complex of 1:1 structure is responsible for signal shifts of lanthanide resonating nuclei. As for the LSR complexes with strong Lewis bases, the lanthanide signal shifts can be caused by 1:1 adduct. The absence of concentration dependence of $t-$Bu signal shifts in the LSR-recorded NMR spectra of these compounds seems obvious. Usually, concentration of S is noticeably higher (or even much higher) than that of the lanthanide. Therefore, due to high values of K_1 and K_2, the position of $t-$Bu signal turns to be is average in rather wide ranges of P values.

1.3. Problems of Hyperfine Coupling Theory and Relaxation Processes in Paramagnetic Complexes with Heterorganic Ligands

In the section 1.2.2, it has been noted that the application of paramagnetic additives is carried out both in "active" and "passive" purposes, i.e. without theoretical calculations of the paramagnetic shift and broadening values. As for the usage of paramagnetic additives in general and LSR in particular, the following explanation should be made.

The HFC manifestation in the NMR spectra is defined by the relaxation processes which belong to the most sophisticated section of the magnetic resonance. In this regard, the remark made in 1965 by Eaton and Phillips is especially spectacular. They have underlined that we poorly understand the relaxation processes in paramagnetic systems and that any chelate or coordination compound of the transition metal can represent a special case, which requires special consideration [5]. Bearing this inference in mind, many researchers aspire to combine the quite in-depth study of the HFC mechanisms and relaxation processes in the paramagnetic complexes with the paramagnetic additives method for the solution of fundamental structural problems. Probably, this is the only approach, the implementation of which could convince the researchers in reliability of structural information obtained from paramagnetic shifts and broadenings. This remark prompts us to consider, at least, the general problems of HFC and relaxation processes (mainly, in paramagnetic complexes with organic heteroatom compounds).

Contact HFC. In the section 1.1.1, it has been stated that the mechanisms of the uncompensated electronic spin delocalization from the coordination ion on and across the ligands can be interpreted in terms of theoretical representations developed still in the sixties of the twentieth century. The key point here is the assumption that HFC transfer (delocalization of the uncompensated electron spin) is realized both at the $\sigma-$, and $\pi-$ systems. From these positions a priori follows that these systems can be considered as independent in the ligand molecules. Such approach, for example, constitutes the basis for the correlation indicating similarity between the perturbation effect of the unpaired electron spin and that of the nucleus spin (section 1.1.1). To confirm the generality of the above similarity, the observed paramagnetic shifts δ in the paramagnetic complex of 1-methylpyrazole and 1-ethylimidazole have been compared with the spin-spin coupling constants J in pyrrole molecule, which is isostructural to pyrazole and imidazole molecules [33].

a b c

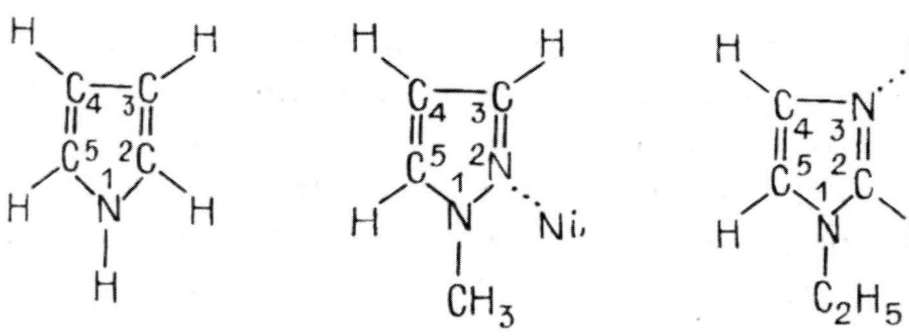

$J_{23} = 2,4 - 3,1$	$\delta_{H-3} = 59$	$\delta_{H-2} = 26$
$J_{24} = 1,3 - 1,5$	$\delta_{H-4} = 56$	$\delta_{H-4} = 36$
$J_{25} = 1,9 - 2,2$	$\delta_{H-5} = 74$	$\delta_{H-5} = 35$
$J_{34} = 3,3 - 3,4$		

Figure 1.4. The constants $\boldsymbol{J}$ (Hz) in pyrrole (a) and $\boldsymbol{\delta}$ values (Hz) in the PMR spectra of 1- methylpyrazole (b) and 1-ethylimidazole (c).

Typical $\boldsymbol{J}$ values between protons in the substituted pyrroles are given in Figure 1.4. The $\boldsymbol{\delta}$ shifts from the region where $\delta = P_M \cdot \boldsymbol{\omega}$ (P_M – is a molar fraction of complex in the sample) are also depicted in Figure 1.4. From the experimental data shown in Figure 1.4 follows that, despite the qualitative comparability, the agreement between contact paramagnetic shifts of pyrazole and imidazole proton signals and $\boldsymbol{J}$ constants in the pyrrole molecule is not sufficiently satisfactory. Indeed, the shifts of H-3 and H-5 proton signals are close, whereas $\boldsymbol{J}_{23}$ value is almost by two times higher than $\boldsymbol{J}_{24}$.Besides, $\boldsymbol{J}_{25} < \boldsymbol{J}_{23}$, though $\boldsymbol{\delta}_5 > \boldsymbol{\delta}_4$. In 1-ethylimidazole, the shifts of H-4 and H-5 protons differ slightly in values, however $\boldsymbol{J}_{35}$ ($\boldsymbol{J}_{24}$) is more than twice lesser than $\boldsymbol{J}_{34}$. Additionally, $\boldsymbol{\delta}_2 < \boldsymbol{\delta}_5$, though $\boldsymbol{J}_{32} >$ $\boldsymbol{J}_{24}$. This discrepancy can be understood if to take into account that the perturbation effect of the nuclear spin is many times weaker than the perturbation effect of the electron spin. Therefore, it is easy to imagine a situation, when the quantitative increase of the perturbation effecting on this system can lead to qualitative alterations in this system that, in its turn, will demand the changes in the approach to assessment of different characteristics of the system and, hence, the corresponding correlations. Thus, a method for HFC mechanisms investigation, based on comparison of $\boldsymbol{\omega}$ (or A) values in the paramagnetic complex with $\boldsymbol{J}$ constants in diamagnetic molecule appears to be approximate.

To achieve the above agreement, theoretical calculations (including quantum-chemical ones) should be performed. The constants $\boldsymbol{J}$ can be applied in these calculations taking into account the fact that it is not necessary to strain after the exact quantitative agreement of their values with the values of paramagnetic shifts. Besides, qualitative agreement of $\boldsymbol{J}$ constants and $\boldsymbol{\omega}$ shifts can be indicative of the realization of one or another mechanism of electron-nuclear coupling transfer in the studied complex.

In this connection, the comparison of paramagnetic shifts in the nitrogen-containing ligands interacting with the central ion by pyridinic nitrogen atom is of certain interest. In 1-methylpyrazole, (Figure 1.4), the ratio $\delta_3/\boldsymbol{\delta}_4 \approx$ 1, and in 1-ethylimidazole, the ratio $\boldsymbol{\delta}_4/\boldsymbol{\delta}_5$ is also almost equal to 1. In pyridine, where $\boldsymbol{\sigma}$-mechanism is realized, the same ratio is 3,92 [18]. Notably that in all three compounds, $\boldsymbol{\omega}$ has identical sign. However, the paramagnetic shifts in the NMR spectra of 1-methylbenzimidazole and 1-methyl-2-aminobenzimidazole (these compounds are coordinated due to pyridine nitrogen atom N-3) turn out to be both low-field and high-field shifted and have different signs [118]. Thus, even the study of HFC mechanism in benzimidazole systems represents an independent task. In even more complex molecules, a possibility of HFC peculiarities manifestation becomes obvious.

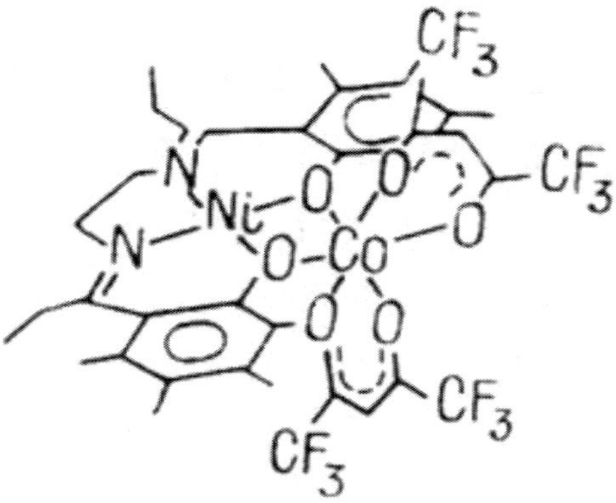

X

For example, all paramagnetic shifts of the aromatic protons in the PMR spectrum of heteronuclear complex **X** are negative (high-field) [119]. Usually, paramagnetic shifts of the aromatic protons are low-field (σ - delocalization mechanism) or have alternating signs (π - delocalization mechanism). Naturally, it is possible to explain the values and signs of the aforementioned shifts by the specifics of the considered systems. For example, it has been assumed [119] that in the studied complexes **X**, the HFC is defined by the indirect σ - mechanism. But this way of qualitative explanation of HFC leads to the situation when peculiarities of such coupling in different complexes come out the view of the researchers.

If the explanation of the HFC nature with participation of ^{1}H nucleus surrounded by only one $1s$ -electron represents a certain problem, it is clear that even more sophisticated problems arise for heavier nuclei surrounded by many electrons. As for the LSR complexes with organic ligands, this problem is also complicated (in comparison with complexes of the first transition group elements) by the fact that $4f$ -orbitals, which electrons induce paramagnetic properties, are effective screened. Therefore, they should not participate in the direct covalent bonding with the ligand orbitals, which main quantum number (in most cases) does not exceed three, and therefore, they significantly differ in energy from $4f$ -orbitals. The nature of paramagnetic shifts and broadenings in the NMR spectra is studied mainly using the resonance from ^{1}H and ^{13}C nuclei. Other nuclei (with small exception) fall out the researcher's view, though modern NMR spectrometers, as we know, in principle allow registering the signals from more than 120 nuclei (including isotopes). As for the complexes of iron group elements, such situation for the atomic nuclei, with which the ligand molecules coordinate with the central ion, becomes natural owing to essential broadenings of the signals. At the same time, it should be noted the following two circumstances:

1) According to the Solomon-Blombergen theory [5], the value $1/T_2$ is directly proportional to γ_N^2 (γ_N – is a gyromagnetic ratio for this nulceus). Therefore, the signal width from other nuclei in comparison with that of the proton signal (other

conditions being equal) is less by $(\gamma_H/\gamma_N)^2$ times (for example, for ^{29}Si – by 25 times, and for ^{15}N – almost by 100 times);

2) As a rule, a number of signals in the NMR spectra of others nuclei (apart from ^{1}H and ^{13}C) is limited, and the range of chemical shifts (in the spectra of diamagnetic organic compounds) is hundreds and thousands ppm.

These two circumstances as well as the progress in NMR spectroscopy devises allow the HFC in paramagnetic systems involving various nuclei to be considered as a promising research direction. In this regard, we should refer to the results of ^{29}Si and ^{119}Sn NMR study of paramagnetic metallocenes XI, which apparently have been carried out for the first time in [120].

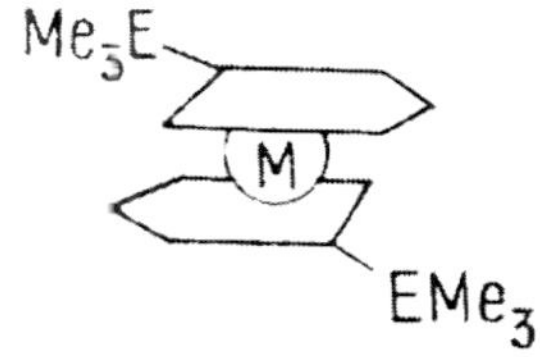

XI

M = V, Cr, Co, Ni;E = Si, Sn

It has been found that the values of paramagnetic shifts for ^{29}Si signals range from – 1319 to 879 ppm. As for ^{119}Sn is concerned, the authors of the work [120] have managed to record the NMR spectrum of only one metallocene (M = V, E = Sn; ω = 755 ppm). It is attempted to rationalize the nature of paramagnetic shifts from positions of contact HFC. This attempt is based on traditional representations (see section 1.1.1). It is pertinent to remember that still in 1967 it has been reported [121] on the results of PMR study of the contact HFC transfer through the E atom (Si, Ge, Sn) in paramagnetic chelates XII of nickel.

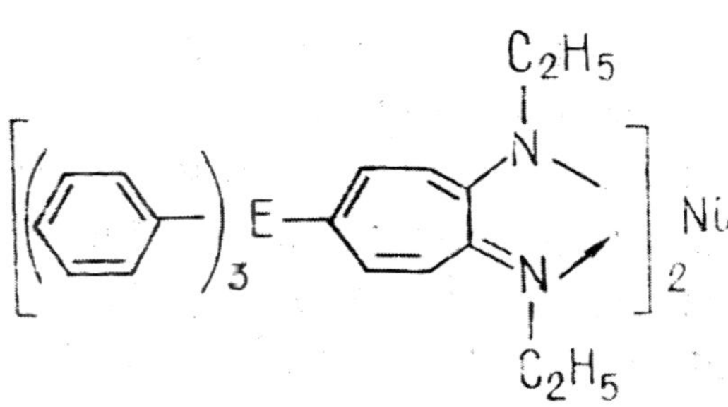

XII

The aim of the study is to elucidate the role of $\mathbf{d}_\pi - \mathbf{p}_\pi$-bonding in the compounds containing E elements in their structure. It should be emphasized that all publications of Eton et al., dedicated to the NMR investigations of the paramagnetic complexes, differ in depth of the problems formulation, coherent explanation and, hence, importance of the received ontained.

It has been reported on the significant contact contribution induced by LSR in the ^{57}Al NMR spectrum of aluminum trisacetylacetone [122]. The authors believe that experimental paramagnetic shifts ω_{Al} are the sum of two contributions (contact and pseudo-contact), which can be separated using the following expression:

$$\omega_{Al} = A\langle S_z \rangle + GD, \quad (1.19)$$

where $G-$ is a geometrical factor equaling to $(3\cos^2\theta - 1)/r^3$. Graphic dependence $\omega_{Al}/\langle S_z \rangle = f(D/\langle S_z \rangle)$ is used for the separation of contact and pseudo-contact contributions. $\langle S_z \rangle$ Values are taken from the work [46], $D-$ values are reported in [123]. It appears that HFC constant A for all employed four lanthanides is positive, however a ratio between components of paramagnetic shifts is essentially different. For example, the contact shift of $Eu(FOD)_3$ is more than two times higher than the pseudo-contact one, whereas $Pr(FOD)_3$ induces the contact shift, which is more than three times less than the pseudo-contact shift, though the value of total paramagnetic shift from both LSR is approximately the same. It is concluded that contact HFC is transmitted due to the spin polarization through the Ln···O···Al fragment. However the mechanism of such a transfer is not discussed.

It has already been mentioned (see section 1.1.1) that the idea about reality of contact contribution for the LSR-induced shifts is originated from the ambition to explain abnormal ω values, which do not fit in the scheme of dipole-dipole (pseudo-contact) HFC. The greatest success in the application of such approach to the explanation of HFC nature has been achieved in the work [46], where paramagnetic shifts of ^{14}N and ^{17}O in lanthanide complexes are theoretically studied. Notably, the ligand molecule coordinates with the central ion through these atomic nuclei. The authors start from the expression (1.4) for contact shift. From this expression it follows that the problem of theoretical calculation of ω_k includes two issues: (i) determination of $\langle S_z \rangle$ value characterizing the paramagnetic complex; (ii) evaluation of A constant which, according to Fermi's formula, "hides" the mechanism of localization of uncompensated electron spin on the resounating nucleus.

First, the authors of the work [46] have calculated $\langle S_z \rangle$ values for all series of trivalent lanthanides with electronic $^n f(n = 1 \div 13)$ configuration, spin-orbital interaction and coupling effects being not separated. Then the calculated values are used to deduce the following dependence

$$\omega = f(\langle S_z \rangle), \quad (1.20)$$

where $\omega-$ are experimental values of paramagnetic shifts for the signals of ^{14}N and ^{17}O nuclei. Unexpectedly, the dependence (1.20) turns out to be of essentially linear character. Therefore, it has been inferred that A constant is negative for ^{14}N (pyridine ligand) and positive for ^{17}O (H_2O ligand), but it does not depend upon the paramagnetic ion nature.

However the delocalization mechanism of uncompensated electron spin in the studied complexes remains obscure though it is concluded that it does not participate in direct covalent bonding of $4f$ -orbitals with the ligand molecule.

The calculated $\langle S_z \rangle$ values [46] turn out to be the best. During more than ten years after publication, they were used by other researchers as parameters for separation of experimental paramagnetic shifts into contact and pseudo-contact components. In 1985, it is reported [124] on the amendment of parameter $\langle S_z \rangle$ value for all trivalent lanthanides (from Ce^{3+} to Yb^{3+}). The fundamental aspect of this study includes the calculation of this parameter for Sm^{3+} and Eu^{3+}.

Here it is pertinent to discuss theoretical aspects of electron-nuclear coupling. The latter is originated owing to the fact that nuclei and electrons have finite sizes. The hyperfine coupling can have both electric and magnetic character. Electric HFC arises between the electric quadruple moment of the nucleus giving $I > 1/2$ spin and a gradient of the field induced by electrons in the location of the nucleus. Magnetic HFC is originated due to the interaction of the nuclear and electric magnetic moments. According to Fermi [125], one can determine a specific type of the Hamiltonian $\hat{\mathcal{H}}_{HFC}$ describing the magnetic HFC between the magnetic moments of nucleus and electron assuming, in particular, that there is a r distance between them. Under the conditions of point dipolar approximation, the HFC Hamiltonian can be expressed as follows (the designations in the formula are commonly accepted):

$$\hat{\mathcal{H}}_{HFC} = 2\beta\gamma\hbar\left[\frac{\vec{I}\left(\vec{L}-\vec{S}\right)}{r^3} + \frac{3\left(\vec{I}\cdot\vec{r}\right)\left(\vec{S}\cdot\vec{r}\right)}{r^5} + \frac{8\pi}{3}\delta(r)\vec{I}\cdot\vec{S}\right]\frac{\mu_0}{4\pi}. \quad (1.21)$$

The first two members of the expression (1.21), given in square brackets, describe usual dipole-dipole (or pseudo-contact) interaction. The latter contains explicit information on the spatial structure of a paramagnetic particle. The third member in this expression corresponds to the interaction between nucleus and s-electron, which is located in the region of the nucleus. This is so-called Fermi contact interaction. It contains information on the delocalization of electron spin on s-orbital of the ligand molecules of a ligand, i.e. it can be considered as a source of information on electron structure of a paramagnetic particle. Generally, the quantitative assessment of this HFC is defined by possibilities of quantum-chemical calculations of the molecules in which the HFC is realized. However, as it was already noted (see section 1 1.1), in some cases, the examination of contact hyperfine coupling independently of the quantum-chemical evaluation can give the valuable results concerning peculiarities of electron structure of a paramagnetic particle.

Currently, there is a plenty of works (and their number continuously grows) devoted to calculations of spin density and the corresponding Fermi contact shifts in complexes of transition metals. Such calculations are performed using various quantum-chemical methods. Naturally, the most exact reproduction of the experiment can be achieved by *ab intio* calculations employing wide basic nuclear orbitals. In particular, *ab intio* calculations of HFC constants on ^1H and ^{17}O nuclei in the $M(H_2O)^{2+}$, ($M = Mn \div Cu$) complexes have been carried out [126]. However even these calculations do not guarantee the reproduction of the correct tendencies in a series of compounds. The method X_α is shown to be perspective for determination of spin densities [127]. Nevertheless, the evaluation of prospects of these methods application should account for the necessity of systematic calculations [128]. It should be stressed that *ab intio* calculations of quite bulky paramagnetic complexes (with high number of atoms in them) are still so far rare. Therefore, the application of semi-empirical methods, in a particular, *INDO* [129] may provide some advantages.

This method proves to be fruitful in quantum-chemical study of structure and HFC mechanisms in paramagnetic complex of 1-vinylimidazole with $NiCl_2$ [130]. Ni – N distance is accepted to be 2,35 Å. The agreement of experimental and theoretical results has been reached assuming the anti-bonding character of the "metal – remote atom" interaction. For the calculation of resonance integrals $H_{\mu\nu}$ of metal nuclear orbitals ($4s,4p,3d$) with all nuclear orbitals, which do not belong to N-3 and Cl atoms, the expression $H_{\mu\nu} = -H^*_{\mu\nu}/R$ is used, where $H^*_{\mu\nu}$ – is the resonance integral calculated by the standard ***INDO*** method; R – is a distance between the corresponding atom and metal atom. The agreement with the experiment is attained when the imidazole rings are rotated by 50° relative to the equatorial plane and the vinyl group is deviated by 45° from the heteroring plane preserving its *trance*-orientation. It has been revealed [130] that the main unpaired spin density is localized on s- and p_x-orbitals of the N-3 nitrogen atom. The p_z-orbital (π-MO of the ligand) contains ~ 0,5% of unpaired spin density, which is localized on the ligand. At the same time, unpaired spin density, localized on s- and p_x-orbitals of the N-3 nitrogen atom, does not take part in further delocalization. Unpaired spin density of other atoms on these orbitals is insignificant.

The complex of 1-vinylimidazole with $CuCl_2$ is calculated in a similar fashion. The studied complex in a solution and in a crystal state has octahedral surroundings of the metal atom. The distance between metal and coordinating nitrogen atom is 2,2 Å. High degree of unpaired electron localization (more than 90%) on the metal orbital is established. The unpaired electron spin density is delocalized from $d_{x^2-y^2}$ metal orbital on the hybrid orbital of lone electron pair of the coordination nitrogen atom. Its further distribution on the ligand molecules proceeds via the direct delocalization across the σ – system and according to spin-polarization mechanism involving with participation of π-systems. Interestingly, the latter mechanism plays very significant role.

Pseudo-contact HFC. According to the current representations, the hamiltonian describing the interaction of the nucleus magnetic moment and unpaired electron with external magnetic field $\vec{H}$ is expressed as follows:

$$\hat{\mathcal{H}} = g\beta\vec{H}\vec{S}_z - g_N\beta_N\vec{H}\vec{I}_z. \qquad (1.22)$$

To describe the NMR phenomenon in paramagnetic system, one should add an effective spin-hamiltonians of (1.21) type, which accounts for the interaction between spins of unpaired electrons and resonating nuclei, to the Zeeman member in the expression (1.22) add. This hamiltonian, which has been discussed in a number of works (see, for example, [131 - 133] and the references therein) comprises two members ($\hat{\mathcal{H}}_{HFC} = \hat{\mathcal{H}}_F + \hat{\mathcal{H}}_D$) characterizing contact and pseudo-contact HFC, respectively. Besides, a member $\mathcal{H}_L = \vec{A}_L \vec{I}$ allowing for the interaction of the nuclear spin with the angular (orbital) moment of the unpaired electron is added to the hamiltonian. Thus, the complete expression can be presented as follows:

$$\hat{\mathcal{H}}_S = g\beta \vec{H} S_z - g_N \beta_N \vec{H} \vec{I}_z + \vec{A}_C \vec{I}, \tag{1.23}$$

where

$$\vec{A}_C = \vec{A}_F + \vec{A}_D + \vec{A}_L. \tag{1.24}$$

The necessity of considering the different mechanisms of the interaction between nuclei and unpaired electrons has stimulated a series of investigations dealing with (i) the alterations of nuclear relaxation times of the NMR signal shifts induced by conductivity electrons (Knight shift); (ii) the polarization of nuclear spins (Overhauser effect); (iii) the paramagnetic shifts caused by isotropic contact and pseudo-contact HPC and other issues.

McConnell and Robertson [57] have shown that the anisotropic magnetic moment of unpaired electrons induces the pseudo-contact shift of the NMR signals in addition to the contact one. Oney and Jesson [134] have deduced the following expression for this shift:

$$\omega_p = \frac{\Delta H}{H} = -\frac{\beta^2 S(S+1)}{3kT} \frac{(3\cos^2\theta - 1)}{R^3} F(g), \tag{1.25}$$

where $F(g)$– is algebraic function of g-tensor components, which depends on a ratio between correlation times characterizing the relaxation of nuclear spins.

In deriving the above equation (1.25) it has been supposed that: 1) R value (distance between the paramagnetic ion and resonating nucleus) should be quite high so that one could consider this ion as a point dipole; 2) the complex has only one thermopopulated energy level (if the magnetic field is absent), and the state of this (main) level can be characterized by spin quantum number S; 3) the splitting of the main level in the absence of magnetic field (splitting in zero field for $S \geq 1$) can be neglected; 4) the orbital contribution H_L to isotropic pseudo-contact shift is taken into account indirectly through the g-tensor components.

The specified limitations are quite severe. Therefore, they could not be neglected always (independently, for example, of the nature of central ion). In this connection, it has been studied [131] the effect of zero field on pseudo-contact (dipole-dipole) shift in the axial-symmetric complexes possessing orbitally non-degenerated states, which are characterized by spin quantum number $S \geq 1$. As an example, we can indicate here the octahedral complexes Ni^{2+}($S = 1$), Cr^{3+}, Mo^{3+} ($S = 3/2$); tetrahedral complexes Co^{2+} ($S = 3/2$); tetrahedral complexes Mn^{2+}, Fe^{2+} ($S = 5/2$). The corresponding analysis leads to the following expressions for pseudo-contact shift taking into account S value [131]:

$$S=1;\omega_p=\frac{2\beta^2\left(g_{II}^2-g_\perp^2\right)}{9kT}\frac{\left(1-3cos^2\theta\right)}{R^3}\left[1-\frac{\left(g_{II}^2+0{,}5g_\perp^2\right)\chi}{3\left(g_{II}^2-g_\perp^2\right)}\right], \quad (1.26)$$

$$S=3/2;\omega_p=\frac{5\beta^2\left(g_{II}^2-g_\perp^2\right)}{12kT}\frac{\left(1-3cos^2\theta\right)}{R^3}\left[1-\frac{4\left(g_{II}^2+\frac{1}{2}g_\perp^2\right)\chi}{5\left(g_{II}^2-g_\perp^2\right)}\right], \quad (1.27)$$

$$S=5/2;\omega_p=\frac{35\beta^2\left(g_{II}^2-g_\perp^2\right)}{36k}\frac{\left(1-3\cos^2\theta\right)}{R^3}\left[1-\frac{224\left(g_{II}^2+\frac{1}{2}g_\perp^2\right)\chi}{\left(g_{II}^2-g_\perp^2\right)}\right], \quad (1.28)$$

where $\chi=D/kT$, $D-$is a splitting parameter in zero field.

The expressions (1.26) – (1.28) have two members which can be divided since the first member defines a contribution from g-tensor anisotropy and depends upon T^{-1}, while the second member, caused by the splitting in zero field, depends upon T^{-2}.

The works of Bleany [132] have covered the fundamental aspects of the NMR study of electron- nuclear interaction nature. In particular, it has been shown that the pseudo-contact shift, caused by trivalent lanthanide shifting reagents, is a consequence of magnetic susceptibility anisotropy of the paramagnetic ion of rare-earth series. If the molecular geometry (as well as number of the electrons on $4f-$orbitals) of the complex in solution remains intact upon varying the temperature of the studied sample, the pseudo-contact shift should change like T^{-2} for all of ions except for $Eu^{3+}$and $Sm^{3+}$and other ions with $4f^5$ and $4f^6$ electronic configuration. For Sm^{3+}, the contribution to the pseudo-contact shift (which depends upon T^{-1}) becomes basic at room temperature (it is more than two times higher than the contribution depending upon T^{-2} and is opposite in sign). For Eu^{3+}, dependence $\omega_p=\omega_p(T)$ has sophisticated character. However in the range of room temperatures, the ω_p value changes significantly faster than T^{-1}.

Like the authors of the work [131], Bleany makes allowance for the fact that the application of the equation (1.25) is conditioned by quite severe limitations, which have already been discussed above. However, the Bleaney's approach appears to be more general. It has been assumed that the field of ligands prevents degeneration of the main state, splitting its energy level by a value which is lesser than kT ($T-$ is absolute temperature).The corresponding hamiltonian accounting for the crystalline field is given by the expression:

$$\mathcal{H}_{kp}=\sum_{k,q}B_k^qQ_k^g, \quad (1.29)$$

where $B_k^q = A_k^q \langle r^k \rangle \langle J \| k \| J \rangle$; J – is a Lande factor; Q_k^q – is a spin operator including J_x, J_y, J_z components of all levels with $k = 2, 4, 6$; $0 \leq q \leq k$; A_k^q – is energy factor; $\langle r^k \rangle$ – is an average radius of $4f$ – orbitals of k – order; $\langle J \| k \| J \rangle$ – is a numerical coefficient. The dependence upon T^{-2} arises only from the operator with $k = 2$ owing to appearance of the following members in the spin-hamiltonian:

$$D_x J_x^2 + D_y J_y^2 + D_z J_z^2 ,$$

with

$$D_x + D_y + D_z = 0. \tag{1.30}$$

Bleany considers only the first nonzero component of the crystalline field operator. Following the theory of pseudo-contact HFC in of lanthanide complexes, the authors of the publication [135] have studied the effect of the crystalline field taking into account the subsequent components (of highest order) of the crystalline field operator. It turns out that the account for components (of higher orders) of the crystalline field operator can lead to about 20% deviation of the paramagnetic shift values calculated by Bleany.

From all the above considerations, it follows that development of the pseudo-contact HFC theory is based on the direct accounting of the interaction between the magnetic moments of unpaired electrons and the magnetic moments of ligands resonating nuclei in dipole-dipole approximation. It has been concluded [136] that this assumption may become correct if the distance between central ion and resonating nuclei is higher than 3,0 Å. Abnormal (from the viewpoint of the HFC theory) paramagnetic shifts are usually considered as a contact contribution. It has been suggested [137] that they can also result from the anisotropy of resonating nuclei shielding tensor. This contribution should be significant for the nuclei, which shielding is defined by Ramsey tensor (for example, for ^{13}C, ^{14}N and other nuclei, but not for protons). In this sense, this contribution is similar to the contact contribution which is caused by polarization of bonds.

Contact mechanism	*Pseudo-contact mechanism*
Delocalization	Anisotropy of g-factor ormagnetic susceptibility
Bond polarization	Anisotropy of shielding tensor

The reality of indirect pseudo-contact HFC makes optional the application of the contact mechanism for explanation of abnormal paramagnetic shifts in the NMR spectra. Besides, indirect contact HFC could be a reason of the constant A invariance as compared to alteration of the central ion in the paramagnetic complex [46]. A possible contribution to the values of paramagnetic shift caused by anisotropy of the shielding tensor $\hat{\sigma}$ via the addition of $\mathcal{H}_A = -\vec{A} \cdot \hat{\sigma} \cdot \vec{I}$ member to the hamiltonian has been theoretically evaluated [138]. It appears that even when σ value is thousands ppm (as, for example, takes place for mercury nuclei), the additional contribution does not exceed one percent from the ω_p value. Probably, H_A does not adequately reflect the mechanism of additional HFC offered in the work [137]. Apparently, its reality will demand additional experimental check involving the analysis of paramagnetic shifts of signals from other, apart from ^{1}H, nuclei, which shielding is defined by the Ramsey tensor.

In this line, the work of Eicher [133] is of fundamental importance for the following reason. From the positions of nature of the electronic structure of multielectronic (molecular) systems, the pseudo-contact shift should be considered as a consequence of either electronic g-factor anisotropy or magnetic susceptibility anisotropy of the paramagnetic ion. Notably, that for lanthanide shifting reagents, the second explanation is more preferable, since it describes the pseudo-contact HFC more correctly [131, 132, 139]. Thus, dipole-dipole shifts can be quite correctly assessed using the values of magnetic susceptibility anisotropy. Besides, the peculiarities of these shifts behavior in the presence of different paramagnetic ions as well as under various experimental conditions (for example, at different temperatures) can be revealed. However Eicher has shown that the description of paramagnetic shifts using the magnetic susceptibility anisotropy (in approximation of point dipole) turns to be incorrect. In his opinion, the works published before the paper [133], incorrectly interpret the interaction of the resonating nucleus spin with orbital motion of unpaired electrons (of the paramagnetic ion). The direct assessment of this interaction shows an additional contribution to the shift

$$\omega_L = -2C_0\left[\frac{(1-k)}{\lambda}\right](3R^3)^{-1}(\chi_{xx} + \chi_{yy} + \chi_{zz}). \qquad (1.31)$$

In this connection, two facts are remarkable. This contribution originates even in the case of realization of spherical symmetry of the central ion environment. The above contribution depends only on R distance, but not on the angular factor. The complete expression for ω_p, derived by Eicher, naturally includes a member with the angular dependence, which is given in the same manner as in the works published prior to [133]. As yet there are no the compelling experimental evidences confirming reality of the contribution defined by the expression (1.41). Moreover, there are no proofs of non- adequacy of the indirect accounting of the orbital contribution $\mathcal{H}_L$ to the pseudo-contact paramagnetic shift through the g-tensor components. It is suggested [133] that the ω_p value should be directly proportional to T^{-1} in the temperature range of 200 – 400 K, provided that the value of energy level splitting (owing to spin-spin coupling) is lower as compared to the kT value. Otherwise, the alterations of paramagnetic shifts upon varying the temperature of the sample will be different.

The experimental study of the magnetic susceptibility dependence upon temperature has shown that the $\omega_p = \omega_p(T)$ function has complex character. Moreover, Horrocks believes that it is unclear whether it is possible at all (at least, in quite wide temperature range) to obtain specific functional dependence on temperature of the paramagnetic shifts induced by lanthanide shifting reagents [140]. It seems to us that Horrocks complicates a problem related to the establishment of explicit dependence of ω_p upon T for LSR. Indeed, the comparative analysis of various theoretical approaches to the assessment of dipole-dipole HFC in the lanthanide paramagnetic complexes demonstrates that, at least, in the temperature range used usually in the NMR studies, one can specify the dependence $\omega_p = \omega_p(T)$ [141,142]. Thus, elucidation of the explicit form of this functional dependence represents one of the key problems of the HFC theory. At the same time, the above dependence is a criterion of adequacy of the applied models for the description of electron-nuclear interaction in real molecular systems. This circumstance together with the problems related to the mechanisms of uncompensated electron spin delocalization stimulates the researches in the field of development of HFC theory in paramagnetic complexes.

Spin nuclear relaxation. According to the existing representations, the system, which is in a non-equilibrium state, eventually will transit into the equilibrium one under certain conditions. Such a transition is currently referred to as the relaxation process. The relaxation phenomena represent irreversible processes. The thermal equilibrium of the spin system with different degrees of freedom can be achieved only through available local magnetic fields. There are some types of the fields which fluctuations are induced by the molecular motion. Among them are the fields of other magnetic nuclei; the fields of unpaired electrons; the fields caused by alternating electron shielding (intra- and intermolecular) of external electrostatic field $\vec{H}_0$. From the positions of nuclear magnetic resonance, the mathematical solution of the task related to the relaxation is reduced to describe movement (over time) of the components M_x, M_y, M_z of macroscopic magnetization $\vec{M}$ of the sample containing N nuclear spins. We mean here the well-known Bloch equations obtained in 1946. These equations contain T_1 and T_2 parameters which characterize the relaxation processes.

The physical nature of these parameters can be understood bearing in mind the existence of different relaxation mechanisms. First of all, this is the spin-lattice interaction which includes all processes of energy exchange between spin system of the resonating nuclei and its environment (for example, crystalline lattice). Generally, all degrees of freedom (except for those of the specific spin system) are referred to as the lattice. The energy transfer from spin system to lattice is characterized by time T_1 which is required for achievement of the energy equilibrium of the system with its environment. The parameter T_1 is also called longitudinal relaxation time since it characterizes the component M_z movement which coincides with the direction of external magnetic field $\vec{H}_0$.

The second type of interaction (spin-spin one) includes all the mechanisms, through which the spins can exchange energy with each other, but not with the lattice as a whole. This is the transverse relaxation, which is characterized by T_2 time and is connected with loss of the spin system coherence. Any process contributing to T_1 will destroy the coherence, therefore, in general case $T_1 \geq T_2$. The physical sense of T_2 can be rationalized using the following model.

Directly after switching off the radio-frequency field inducing the transitions between energy levels, individual spins begin to process around the field $\vec{H}_0$. In the absence of a relaxation, all characteristics of the spins motion would rotate with the same frequency. Therefore, the starting values of M_x and M_y parameters would also remain constant in time. Different casual interactions of the magnetic dipoles induce the fluctuating magnetic fields amplifying or weakening $\vec{H}_0$ and thus changing the precession rate of separate spins. As a result, the spins lose synchronism of the rotation and their distribution "is smeared" in time over more and more broader region in the xy plane. Eventually, the resultant transverse (relative to field $\vec{H}_0$) component $\vec{M}$ turns to zero. Thus, T_2 value characterizes time, over which the coherence of spin motion completely disappears.

Paramagnetic ions can cause the considerable reduction of relaxation times T_1 and T_2 of nuclear spins due to strong oscillatory magnetic fields ($\vec{H}_{loc}$) generated by unpaired electrons, which magnetic moment is by three orders higher than the maximum nuclear magnetic moments. Thus, [143]:

$$\vec{H}_{loc} \approx Sp\{\mathcal{H}^2\}.$$

Pioneer researches in this direction were conducted by the authors of the works [144,145] who obtained the expressions (now they are referred to as Solomon – Blombergen equations) describing the relaxation times T_1 and T_2 of the nuclei interacting with paramagnetic ions. Usually, the so-called Solomon – Blombergen modified equations are used [126] (analysis of their applicability to the method of complex-forming paramagnetic additives is given in section 2.1 of this monograph). The application of these equations is restricted by following conditions.

1. Electronic spin is considered as a point dipole centered on the metal ion.
2. The electronic g-factor should be isotropic.
3. Molecular fluctuations should be quite fast to average any anisotropic part of spin-hamiltonian.
4. Correlation function of electron-nuclear interaction should be exponential correlation with the only correlation time τ_r.
5. It is considered that Redfield ratio is valid for electron spin relaxation.
6. Chemical exchange is not correlated with other fluctuations in the lattice.
7. Fluctuations in the spin-electronic system do not correlate with oscillations in other parts of the lattice.
8. The spin-electronic relaxation is characterized by one-exponential function.

The first two conditions relate to the intensity of electron-nuclear interaction. The second assumption was discussed in detail by Sternlicht. He has found that anisotropy of the g-factor under certain conditions can influence the rate of the nuclear relaxation. Conditions 4 – 8 are connected with the dynamics of electron-nuclear interaction. In particular, the fourth condition means that internal fluctuation in a paramagnetic complex and molecular reorientation are isotopic with the correlation time $\tau_r = 1/6D$, where $D-$ is rotary-diffusive coefficient. In principle, both the internal rotation and anisotropic diffusion can be taken into account in the method developed by Vossner. The eighth condition is a consequence of the fact that the Solomon – Blombergen equations are not quite strict in the case, when spin of paramagnetic particle is $S \geq 3/2$. In this case, the relaxation of electron is multiexponential and characterized by $(S+1/2)$ or S exponents for half-integer and integer values of S, respectively. This question has been studied by Rubinshtein et al. The condition 7 means that the physical processes related to the electronic and nuclear spin-lattice relaxation should not correlate with each other. If, however, the electronic spin system satisfies to the Redfield condition, this restriction disappears. The detail analysis of the above conditions is given in [146 – 149]. It should be noted that the application of the Solomon – Blombergen equations is based on the assumption that the molecules, the NMR spectra of which are analyzed, enter and leave the coordination sphere of the paramagnetic in a random way [5]. Usually, it is also supposed that only one stable complex exists in solution. Generally, a character of the relaxation will be defined also by exchange processes, which should be taken into account together with nature of the paramagnetic ion and the experimental conditions (see section 2.1).

However, there are two general cases of the nuclear spin relaxation in the presence of HFC. The first one is realized under the conditions of short times of electronic relaxation, when $T_1 = T_2$. In this case, the NMR signals broaden insignificantly. The second one is realized when relaxation times of the paramagnetic ion are quite high. In this case, $T_1 >> T_2$ and the NMR lines broaden so significantly that often it is impossible to separate them. This situation turns to be typical for ions in the s-state, which are characterized by the electronic relaxation time equaling to $10^{-8} - 10^{-9}$ s. Organic radicals may have even higher electronic relaxation times [5]. Up to the sixties of the last century, this was the reason hampering the detection of NMR signals from these paramagnetic particles in solution.

Starting point of the Solomon – Blombergen theory is an idea of the interaction between unpaired electrons centered on the coordinating ion and resonating nuclei of the ligands in approximation of the point dipole. It means that the following ratio takes place:

$$T^{-1} \sim R^{-6} f(\tau), \tag{1.32}$$

where $T-$ is a nucleus relaxation time (longitudinal or transverse); $R-$ is a distance from the nucleus to the coordination ion; $f(\tau)-$ is a function, explicit form of which is defined, in particular, by a ratio between the correlation times characterizing electron- nuclear interaction in the system. The finitude of the nuclear orbitals occupied by unpaired electrons prompts the researchers to elucidate the limits of applicability of the point dipole approximation (see [126, 150,151 and the references given there). It is offered to replace a cofactor R^{-6} in the expression (1.32) by R_{eff}^{-6}, which is determined as a value of the expected parameter calculated using the wave function of the system (complex).

Here it is pertinent to mention the work [126], the authors of which have offered a test for checking applicability of the point dipole approximation. The test is based on *ab intio* calculations of paramagnetic complexes $M(H_2O)_6^{2+}$ ($M = Mn, Fe, Co, Ni, Cu$). It appears that this approximation quite satisfactorily describes the internuclear distance in the case of hydrogen atoms. At the same time, the $\boldsymbol{R}$ value between oxygen and the coordination ion, found in this approximation, is less than the expected one. It is known that the performance of quantum-chemical calculations (especially *ab initio* ones) of real paramagnetic complexes is rather complex problem which solution a priori is not always connected with obtaining adequate information on the molecular structure. This is why the researches should apply the expression (1.32) in their current experiments. It is shown [136] that when $\boldsymbol{R} \geq 7$ Å the dipole-dipole approximation for the description of HFC between unpaired f-electrons and nuclei is quite correct (within the limits of experimental error). However, in the complexes with unpaired electrons on d-orbitals, the limitation of the above approximation is observed up to value of $\boldsymbol{R} \cong 7$ Å [152]. Thus, it is possible to assume that application of the dipole-dipole member of the Solomon – Blombergen equations is an impelled step, but in a number of cases, it becomes reasonable.

The NMR studies of the relaxation processes in paramagnetic complexes are aimed also at the establishment of implicit form of $f(\tau)$ function in the expression (1.32). Especially, it relates to the ions having quite long relaxation times of electronic spins ($Cu^{2+}, Mn^{2+}, Gd^{3+}$) and when $\boldsymbol{S} > \mathbf{1/2}$. Without going into details of this problem, which should be discussed elsewhere, it should be emphasized the following fact [153]. In the context of structural studies, unambiguity of the conclusions drawn on the basis of the NMR spectra transformed by HFC, assumes a selection of paramagnetic ion and conditions of the spectra recording, which would be appropriate for the implementation of the so-called fast exchange (see section 2.1).

Chapter 2

METHOD OF PARAMAGNETIC ADDITIVES

Currently it is a common knowledge that the nuclear magnetic resonance represents a potent tool for the study of multielectron systems since parameters of the NMR spectra contain valuable information on electronic and spatial structure of the molecules. At the same time, the assignment of resonance lines in the NMR spectra using the simple rules of the analysis is not always unambiguous. The creation of NMR spectrometers operating at resonance frequencies of 400 MHz, 500 MHz and higher provides ample opportunities for manipulations with detected spectral information, thus expanding a range of compounds, spectral analysis of which (by the above-mentioned rules) is still effective. However, the new tasks arise in this direction. They relate to the investigations of molecular structure of multielectron systems which cannot be studied by the above rules only.

From the chapter 1 of this monograph follows that upon coordination of a molecule with paramagnetic ion, the resonance lines may have essentially different paramagnetic shifts (contact and pseudo-contact ones), i.e. they can be shifted differently relative to the position in the spectrum of free ligand. As a consequence, the difference in chemical shifts of the resonating nuclei' signals becomes higher than the constant of spin-spin coupling between the same nuclei (first of all, between protons). In the result, spectral analysis will be significantly simpler. Apparently, this circumstance has been used for the first time for determination of spin-spin coupling constants in the substituted aminoprotonoiminates XIII coordinated with nickel. Detailed discussion of this issue is given in the work [154]. It is assumed that spin-spin coupling in aromatic and unsaturated compounds is transferred (across not more than three bonds) predominantly by $\sigma-$electrons. In the paramagnetic complexes of nickel with compounds XIII, the contact paramagnetic shifts are caused by perturbation of the $\pi-$system of the ligand molecule. Therefore, it can be expected that the J value, determined from the spectra of proton magnetic resonance (PMR) of the paramagnetic complex, will be close to J values in diamagnetic molecule.

The ligand has the following structure:

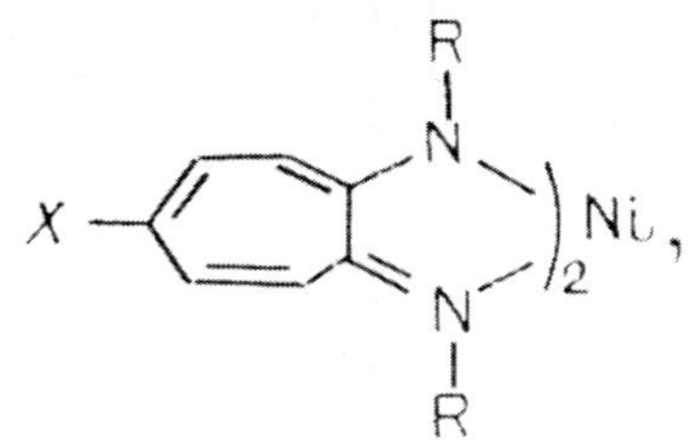

XIII

where $X = H$ or $\boldsymbol{Br}$; $\boldsymbol{R}$ – are various aliphatic or aromatic groups.

In Figure 2.1, a fragment of the PMR spectrum of the paramagnetic complex (curve 2) is shown, when:

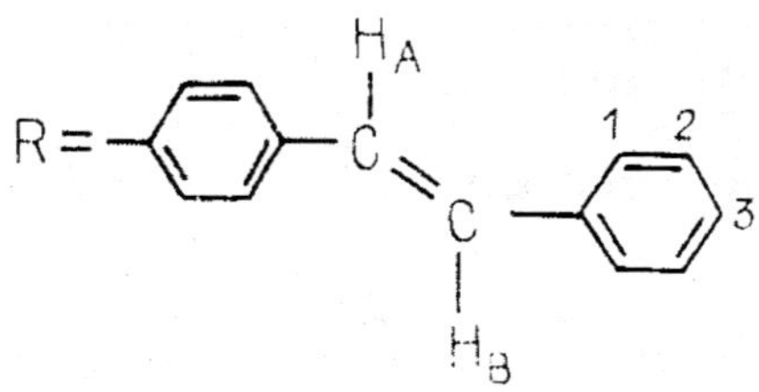

For comparison purposes, the Figure 2.1 shows the PMR spectrum of *trans*-stilbene taken from catalogue of Varian company.

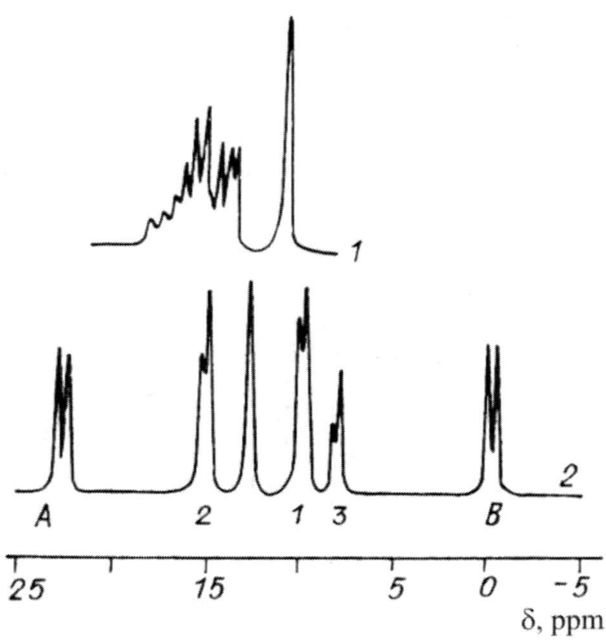

Figure 2.1.The PMR spectrum of *trans*-stilbene (1) and a fragment of the PMR spectrum (2) caused by protons of *trans*-stilbene.

In the *trans*-stilbene, olefinic protons are chemically nonequivalent, so that spin- spin coupling between them is not observed, and their signal appears as a singlet ($\delta = 7, 10$ ppm) in the PMR spectrum. Aromatic protons are responsible for the appearance of unresolved multiplet in the spectrum. The resonating nuclei of the same molecule in the paramagnetic complex become significantly nonequivalent (see Figure 2.1, curve 2). Nickel complexes with compound XIII in solution are characterized by fast intramolecular equilibrium of the following type:

$$\begin{array}{ccc} \text{Diamagnetic form} & \rightleftarrows & \text{Paramagnetic form} \\ \text{(square)} & & \text{(tetrahedron)} \end{array}$$

This leads to narrowing of the resonance lines in the PMR spectra of the complexes that allows the spin-spin coupling with $J = 2-3$ Hz to be detected.

2.1. Theory of the Method

As it has already been noted, very often it is impossible to record the high resolution NMR spectrum of the paramagnetic complex due to considerable broadening of the resonance lines. This especially relates to the NMR resonance signals from carbon, nitrogen and some other atoms, which constitute skeleton of the molecules. For example, broadenings of signals from protons can reach hundreds and even thousands hertz. As for the signals of carbon atoms, they can broaden so significantly that their detection becomes problematic. Figure 2.2 and 2.3 depict the ^{13}C NMR spectra of 1-vinylimidazole (VIM) and its complex with $CoCl_2$, $(VIM)_4CoCl_2$, recorded atroom temperature ($CDCl_3$ as a solvent). The spectrum of VIM is recorded with proton decoupling; the spectrum of VIM-cobalt complex is recorded without proton decoupling [155]. Here and in other figures, the solvent signals are designated by the asterisk. As is expected, the spectrum shown in Figure 2.3 contains five signals. The signal of the vinyl group atom C(α) is present as a doublet owing to spin-spin coupling with H(X) proton, while signal of the C(β) atom appears as a triplet due to spin-spin coupling with H(A) and H(B) protons. Carbon atoms of the imidazole cycle cause the emergence of three strongly broaden signals, which are also significantly shifted relative to their positions in the ^{13}C NMR spectrum of non-coordinated VIM (see Figure 2.2). In the spectrum of VIM-nickel complex, only signals of the vinyl group atoms (i.e. C(α) and C(β)) are observed, whereas signals of carbon atoms of the imidazole ring are not detected owing to their strong paramagnetic broadening. For this reason, in the ^{13}C NMR spectrum of the copper complex, the signals from carbon atoms are not detected at all.

The aforementioned spectral peculiarities are explained by the fact that the relaxation times of electron spin in complexes of cobalt, nickel and copper are significantly different. The cobalt complex has the shortest relaxation time (approximately 10^{-12} s), while for the copper complex, this value is considerably higher (~ 10^{-9} s).The shorter the relaxation time, the shorter will be the contact of resonating nuclei with uncompensated electron spins. It is natural that eventually transformation of the NMR spectrum will be less expressed, and this is observed in the spectra discussed here. It should be added that among the elements of iron subgroup, the paramagnetic ion Co^{2+} can be used as a shifting reagent.

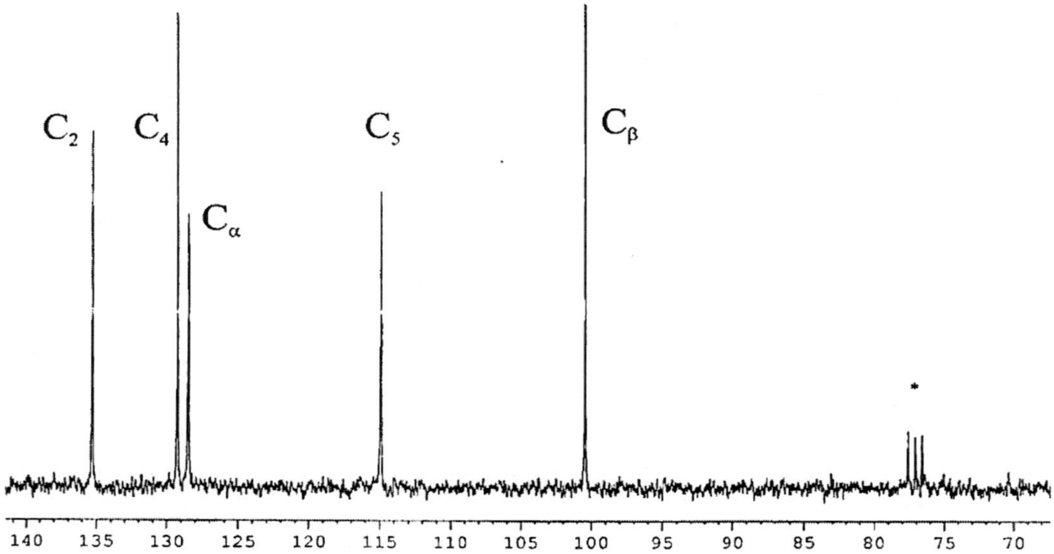

Figure 2.2. ^{13}C NMR spectrum of 1-vinylimidazole.

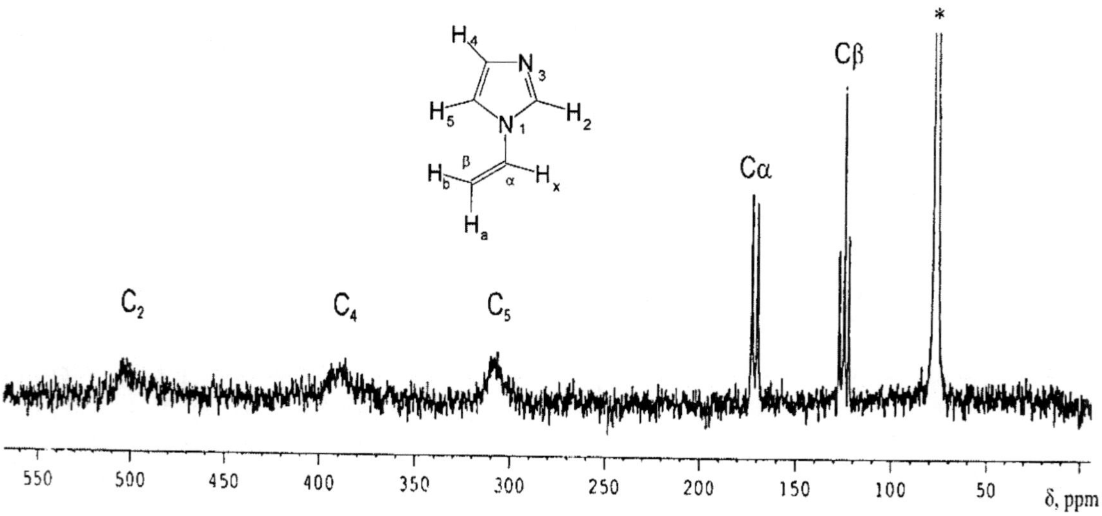

Figure 2.3. ^{13}C NMR spectrum of $(VIM)_4CoCl_2$ complex.

The difficulties caused by paramagnetic broadening can be circumvented if to record the spectra of samples, in which the metal – ligand ratio is much less than 1. Due to the fast molecular exchange between the coordination sphere of the complex and free ligand, the observed spectrum becomes average. The shape of the spectrum will depend, in particular, on nature of electron- nuclear interaction in the paramagnetic complex. This approach is based on accounting for the effects of exchange processes on shape of the NMR spectrum. It is a common knowledge that if proton H participates in the fast exchange of $\boldsymbol{H(A)+B\rightarrow}$

$H(B)+A$ type, the observed shift of its signal is defined by the expression $\delta_H = P_A\delta_A + P_B\delta_B$, where P_A and P_B – are molar fractions; δ_A and δ_B – are chemical shifts of the proton signal in environment of A and B. In our case:

$$\delta_H = P_C\delta_C + P_M\delta_M = \delta_C + (\delta_M - \delta_C)P_M = \delta_C + \omega P_M, \quad (2.1)$$
$$\delta_M - \delta_C = \omega,\ P_C + P_M = 1,$$

where the interlinear index C relates to free molecule, and the index M – relates to the molecule coordinated with paramagnetic ion.

For assignment of resonance lines in the spectra of the samples containing paramagnetic complexes, only one complex should be present in the solution upon varying the metal/ ligand ratio. It is easy to see that linear dependence of δ_H upon P_M is a criterion of complex formation in the studied sample. Besides, the expression (2.1) allows the values δ_M (when $P_M = 1$) and δ_C (when $P_M = 0$) to be determined via the extrapolation of graphic dependence $\delta_H = f(P_M)$. Such approach has been applied for assignment of signals in the PMR spectra of pyridine,- quinoline- and isoquinoline-N-oxide [156].

It has already been mentioned that the additives of paramagnetic complexes are used, as a rule, in the presence of excess concentration of ligand in the sample. Owing to the molecular exchange between the coordination sphere of paramagnetic complex and free ligand, the NMR spectrum becomes average. The quantitative assessment of shifts and broadenings of the resonance lines (because of the hyperfine coupling in this case) is reduced to the solution of the Bloch equations, modified by McConnell who has introduced a chemical exchange in them [157]. The detail investigation of the effect of molecular exchange between the coordination sphere of paramagnetic complex and free (non-coordinated) ligand in solution on the width and positions of the NMR signals has been carried out by authors of the publication [158], which results constitute a basis of a method for determination of functional groups position in molecules of heteroatom organic compounds [95, 159 – 161]. Further research in this direction is performed by the authors of the work [162]. The case when the studied sample contains a complex of one structure has been considered. The analysis of the Bloch – McConnell equations leads to the following expressions:

$$\Delta = \frac{P_M}{\tau_M}\left[\frac{(1/T_{2M})^2 + 1/T_{2M}\cdot\tau_M + \omega^2}{(1/\mathrm{T}_{2M} + 1/\tau_M)^2 + \omega^2}\right], \quad (2.2)$$

$$\delta' = \delta'_{\kappa} + \delta'_{\mathrm{p}} = \frac{P_M\ \omega}{\tau_M^2\left[(1/T_{2M} + 1/\tau_M)^2 + \omega^2\right]}, \quad (2.3)$$

where τ_M – is a life time of the ligand molecule in the first coordination sphere of the complex; P_M – is a molar fraction of the complex in the sample; $\omega = \omega_{\kappa} + \omega_{\mathrm{p}}$ – is a shift of the resonance line of the nucleus, contained in the paramagnetic complex, caused by contact ω_K and pseudo-contact ω_p components. Here it should be underlined that in a number of works, it is reported that when nickel acts as a coordination atom, the values of the pseudo-contact shift are insignificant.

In the equations (2.2) and (2.3), T_{2M} – is a transverse relaxation time of resonating nucleus in the complex, which is expressed by the formula (Solomon – Blombergen expression) [163]:

$$1/NT_{2M} = (1/15)S(S+1)g^2\beta^2\gamma_N^2(n/N_p)r^{-6}\{7\tau_c + 13\tau_c/(1+\omega_s^2\tau_c^2)\}+ \\ +(1/3)S(S+1)A^2\hbar^2(n/N_p)\{\tau_e + \tau_e/(1+\omega_s^2\tau_e^2)\}. \quad (2.4)$$

The similar formula describes the longitudinal relaxation time in the complex [163]:

$$1/NT_{1M} = (2/15)S(S+1)g^2\beta^2\gamma_N^2(n/N_p)r^{-6}\{3\tau_c + 7\tau_c/(1+\omega_s^2\tau_c^2)\}+ \\ +(2/3)S(S+1)A^2\hbar^2(n/N_p)(1/\omega_s^2\tau_e^2), \quad (2.5)$$

where n – is a number of nuclei in the paramagnetic complex; N_p – is a molar concentration of resonating nuclei; N – is a molar concentration of paramagnetic ions; ω_s – is a frequency of Larmor electron precession; τ_e and τ_c – are correlation times of contact and pseudo-contact interaction, respectively:

$$1/\tau_e = 1/\tau_s + 1/\tau_M\,;\ 1/\tau_c = 1/\tau_s + 1/\tau_r + 1/\tau_M\,. \quad (2.6)$$

where τ_s – is a relaxation time of electron spin; τ_r – is a reorientation time of the paramagnetic particle. Since τ_e and τ_c are characteristics of random processes, each specific situation requires a special consideration.

Very often τ_e^{-1} and τ_c^{-1} can differ considerably from ω_s, thus predetermining the contributions of pseudo-contact and contact interaction to T_{1M} and T_{2M}. For example, for Mn^{2+} at a frequency of the proton resonance of 30 MHz, the following inequalities are valid [164]:

$$\omega_s \cdot \tau_e >> 1, \omega_s \cdot \tau_c << 1. \quad (2.7)$$

For the inequalities (2.7), the expression (2.4) and (2.5) are replaced as follows:

$$1/NT_{1M} = (4/3)S(S+1)g^2\beta^2\gamma_N^2(n/N_p)r^{-6}\tau_c, \quad (2.8)$$

$$1/NT_{2M} = (4/3)S(S+1)g^2\beta^2\gamma_N^2(n/N_p)r^{-6}\tau_c + \\ +(1/3)S(S+1)A^2\hbar^2(n/N_p)\tau_e. \tag{2.9}$$

Hence, the contact interaction (second member in the last equation) will contribute to the line width. The contribution can be decreases, if

$$\omega_s \cdot \tau_e << 1, \omega_s \cdot \tau_c << 1. \tag{2.10}$$

In this case,

$$1/NT_{1M} = (4/3)S(S+1)g^2\beta^2\gamma_N^2(n/N_p)r^{-6}\tau_c + \\ -(2/3)S(S+1)A^2\hbar^2(n/N_p)\tau_e, \tag{2.11}$$

$$1/NT_{2M} = (4/3)S(S+1)g^2\beta^2\gamma_N^2(n/N_p)r^{-6}\tau_c + \\ +(1/3)S(S+1)A^2\hbar^2(n/N_p)\tau_e, \tag{2.12}$$

i.e. $1/T_{1M} = 1/T_{2M}$. The latter expression for $1/T_{2M}$, like the former, contains the second member, but its contribution is significantly lesser, since τ_e in the expression (2.12) is much lesser than τ_e in the expression (2.9).The inequalities (2.10) are realized when τ_c is equal to τ_e or when the values of these correlation times differ slightly. This is correct when the contact and pseudo-contact interactions are defined by the same characteristic time, introduced in the both expressions (2.6).Indeed, when the frequency of proton resonance is 100 MHz, $\omega_s \cdot \tau_e << 1$, if τ_e equals 10^{-12} s. Such value corresponds to the τ_s values of the elements which are characterized by very short relaxation times of electron spin. Besides, an essential strengthening of the inequality $\omega_s \cdot \tau_e << 1$ (i.e. narrowing of the resonance lines) will take place when temperature of the studied sample increases. This is a consequence of the fact the Eyring formula, used for determination of τ_M, contains temperature inexponent.

If to apply the formula (1.5) for $\omega_ê$, the expression (2.12) can be presented as follows:

$$1/NT_{2M} = (4/3)S(S+1)g^2\beta^2\gamma_N^2(n/N_p)r^{-6}\tau_c + \\ +\{(6k^2T^2)/\hbar^2 g^2S(S+1)\}(n/N_p)(\gamma_N/\gamma_e)^2\tau_e\omega_\kappa^2, \tag{2.13}$$

or

$$1/T_{2M} = Br^{-6} + C(\delta_k')^2, \tag{2.14}$$

where δ'_k – is the observed contact shift of the resonance line; r – is a distance from the coordination ion to resonating nucleus; B and C – are constants for the studied sample at given temperature. Thus, for the complexes, in the spectra of which, paramagnetic shifts are pseudo-contact, the broadenings of signals are defined by the expression

$$1/T_{2M} = Br^{-6}. \quad (2.15)$$

It is obvious that depending on the ratio between τ_M, T_{2M} and ω, the expressions (2.2) and (2.3) can be simplified. The ratios between these values are defined by nature of the studied complexes as well as by the experimental conditions. It is not difficult to see that in the following cases, the expression (2.2) and (2.3) are replaced by simpler ones [158, 162]:

$$\text{slow exchange: } \Delta = P_M/\tau_M, \quad (2.16)$$

$$\delta' = -P_M\omega T_{2M}^2/\tau_M^2 \text{ or } \delta' = -P_M/\tau_M^2\,\omega; \quad (2.17)$$

$$\text{fast exchange: } \Delta = P_M\tau_M\omega^2 \quad (2.18)$$

$$\text{or } \Delta = P_M/T_{2M}, \quad (2.19)$$

$$\delta' = -P_M\omega; \quad (2.20)$$

$$\text{intermediate case: } \Delta = P_M/T_{2M} + P_M\tau_M\omega^2, \quad (2.21)$$

$$\delta' = -P_M\omega. \quad (2.22)$$

It should be underlined that Δ and δ' are functions of the same variables. Therefore, depending on nature of the complexes and experimental conditions, one can use either both expressions (2.2) and (2.3) or one of them.

The conditions of slow exchange are realized in two cases:

$$\omega^2 >> 1/\tau_M^2, \omega^2 >> 1/T_{2M}^2 \quad (2.23a)$$

or

$$1/T_{2M}^2 >> 1/\tau_M^2, 1/T_{2M}^2 >> \omega^2. \quad (2.23b)$$

The alteration of Δ and δ' upon increase of the sample temperature will depend on which inequality (2.23) is realized in the case of slow exchange. This assumption is supported by the data given in Table 2.1.

Δ and δ' values are calculated taking into account the fact that upon increase of temperature, the values $1/T_{2M} \cdot \tau_M$ becomes comparable with ω^2 (2.23a) or with $1/T_{2M}$ (2.23b). Under these conditions, the expression (2.2) and (2.3) can be presented as follows:

$$\Delta = \frac{P_M}{\tau_M}\left(\frac{1/T_{2M}\tau_M + \omega^2}{2/T_{2M}\tau_M + 1/\tau_M^2 + \omega^2}\right); \quad \delta' = -\frac{P_M\omega}{\tau_M^2\left(2/T_{2M}\tau_M + 1/\tau_M^2 + \omega^2\right)};$$

for inequality (2.23b)

$$\Delta = \frac{P_M}{\tau_M}\left(\frac{1/T_{2M}^2 + 1/T_{2M}\tau_M}{1/T_{2M} + 2/T_{2M}\tau_M + 1/\tau_M^2}\right); \quad \delta' = -\frac{P_M\omega}{\tau_M^2\left(1/T_{2M}^2 + 2/T_{2M}\tau_M + 1/\tau_M^2\right)}.$$

To specify Δ and δ' values, it is assumed that $T_{2M} \cdot \omega$ = 10. Such condition corresponds to the inequality (2.23a).

Table 2.1. Dependence of δ' and Δ values upon τ_M

Inequality	τ_M	δ'	Δ	$\lvert\delta'\rvert/\Delta$
2.23a 2.23б	$2/\omega$ $2T_{2M}$	$0{,}80\,P_M/\tau_M^2 = 0{,}20P_M\omega$ $0{,}44\,P_M\omega T_{2M}^2/\tau_M^2 = 0{,}11P_M\omega$	$0{,}80\,P_M/\tau_M = 0{,}20P_M\tau_M\omega^2$ $0{,}67\,P_M/\tau_M = 0{,}33\,P_M/T_{2M}$	0,50 3,28
2.23a 2.23б	$1/\omega$ T_{2M}	$0{,}50\,P_M/\tau_M^2 = 0{,}50P_M\omega$ $0{,}25\,P_M\omega T_{2M}^2/\tau_M^2 = 0{,}25P_M\omega$	$0{,}50\,P_M/\tau_M = 0{,}50P_M\tau_M\omega^2$ $0{,}50\,P_M/\tau_M = 0{,}50\,P_M/T_{2M}$	1,00 5,00
2.23a 2.23б	$1/2\omega$ $T_{2M}/2$	$0{,}20\,P_M/\tau_M^2 = 0{,}80P_M\omega$ $0{,}11P_M\omega T_{2M}^2/\tau_M^2 = 0{,}44P_M\omega$	$0{,}22\,P_M/\tau_M = 0{,}88P_M\tau_M\omega^2$ $0{,}33\,P_M/\tau_M = 0{,}66\,P_M/T_{2M}$	1,81 6,67
2.23a 2.23б	$1/3\omega$ $T_{2M}/3$	$0{,}10\,P_M/\tau_M^2 = 0{,}90P_M\omega$ $0{,}06\,P_M\omega T_{2M}^2/\tau_M^2 = 0{,}54P_M\omega$	$0{,}12\,P_M/\tau_M = 1{,}08P_M\tau_M\omega^2$ $0{,}25\,P_M/\tau_M = 0{,}75\,P_M/T_{2M}$	2,50 7,20
2.23a 2.23б	$1/4\omega$ $T_{2M}/4$	$0{,}06\,P_M/\tau_M^2 = 0{,}96P_M\omega$ $0{,}04\,P_M\omega T_{2M}^2/\tau_M^2 = 0{,}64P_M\omega$	$0{,}08\,P_M/\tau_M = 1{,}28P_M\tau_M\omega^2$ $0{,}20\,P_M/\tau_M = 0{,}80\,P_M/T_{2M}$	3,00 8,00

The data shown in Table 2.1 show that change of temperature appears to be less effective, if in the slow exchange, Δ and δ' values are defined by the relaxation of nuclear spin in the complex. Besides, the fast exchange for δ' is reached faster than for Δ. For example, when τ_M changes from $2/\omega$ to $1/3\omega$ (2.23a) and from $2T_{2M}$ to $T_{2M}/3$ (2.23b), the δ' value increases by 4,5 and 5,5 times, respectively, whereas the broadening of the line decreases by

1,1 and 2,3 times. It should be noted that $|\delta'|$ can be significantly higher than Δ when Δ differs slightly from P_M/τ_M value.

It has already been mentioned, for intermediate case (upon transition from slow exchange to fast and vice versa), the Δ value is defined by the expression (2.21). The temperature range applicable for this expression depends on the inequality (2.23a or 2.23b) which will be used for determination of the slow exchange conditions. The majority of lanthanide complexes is characterized by very short times of electron spin relaxation [1]. Therefore, in the spectra of samples containing lanthanide shifting reagents, the condition of slow exchange should be realized at significantly lower temperatures as compared with complexes of iron group elements.

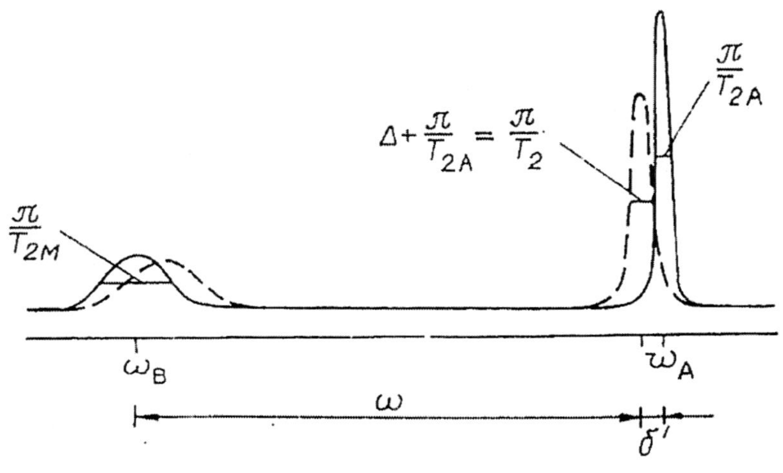

Figure 2.4. NMR signal of resonating nuclei for non-coordinated (ω_A) and coordinated (ω_B) state (dot line corresponds to chemical exchange).

Figure 2.4 illustrates the effect of chemical exchange on the position and width of resonance lines in the NMR spectra [162]. The dotted lines correspond to the case of chemical exchange. Since T_{2M}, τ_M, ω are the functions of temperature T, simplification of the expressions (2.2) and (2.3) is connected with the following

$$\Delta = f(T), \tag{2.24}$$

$$\delta' = f(T). \tag{2.25}$$

The diagram of such dependence can be found in the work [162].

If to mean possibility of obtaining the structural information, the expressions for the observed Δ and δ', corresponding to the case of fast exchange, are the most convenient. The absence of complete averaging is supported by the dependence of the δ' value (expressed in ppm) on the resonance frequency of spectrometer [165]. A method for verification of the fast exchange conditions has been proposed in the work [166]. It consists in the comparison of the observed ratio Δ/δ' with the expected ratio between broadening and shift for the fast exchange. However, the application of Cu^{2+}-induced broadenings in structural researches should be made with care [167 – 169]. This paramagnetic ion is characterized by quite high τ_s value. As a consequence, often the fast exchange will not be realized at a temperature which significantly exceeds room one. This makes it difficult to evaluate the contribution of first member of the expression (2.14), depending on relative positions of resonating ligand nuclei and coordination ion.

Thus, paramagnetic shifts and broadenings have the characteristic temperature dependence, the study of which allows finding the thermodynamic parameters of complex formation in solution as well as elucidating stoichiometry of complexes in solution and mechanisms of ligand exchange. In the reviews [107, 170, 171], one can find the relevant references covering both the experimental protocols and possibilities of the method.

2.2. Paramagnetic Additives of Iron Group Elements

The method described in paragraph 2.1 has been applied for the first time for interpretation of the PMR spectra of pyrazole methyl derivatives as well as 1-vinyl- and 1-ethylimidazole [95, 159]. The proton signals in the spectra of VIM and EIM have been assigned by evaluation of broadenings and shifts of the resonance line signals. The experiments are carried out in the presence of excess concentration of ligand in the sample.

Figure 2.5 shows the temperature dependence of proton signals broadening in the PMR spectra of VIM recorded with additive of the paramagnetic complex $NiCl_2{\cdot}4VIM$. It is seen that resonance lines of two protons in the imidazole cycle broaden identically up to 303 K. The resonance line of the ring third proton approximately up to 323 K is overlapped with the line of $-CH-$proton of the vinyl group. However, shape of this signal allows one to conclude that its resonance line up to 303 K is defined by the expression (2.16). Starting from 313 K, two lines are narrowed approximately in the same manner, while the third one narrows faster than others. This difference increases approximately up to 353 K and then remains constant. At a temperature over 353 K, Δ value of one of the lines will be by 10 times lesser than Δ value of two others.

It is easy to verify that beginning from 313 K, the broadening of the ring proton signals cannot be defined by the expression (2.18). Indeed, in this case, shifts of these proton signals would be defined by the expression (2.20). But the observed δ' values for three resonance lines differ slightly (Table 2.2). Therefore, for three signals, δ' values (and hence Δ values) should be similar. However, the dependence (2.24) does not confirm this (see Figure 2.5).

To establish a type of approximation in formula (2.2) in the temperature region of 313 – 370 K, it is supposed [159] that the expression (2.14) for T_{2M} consists of two members, which depend upon r^{-6} and $(\delta')^2$, respectively. The $(r_5/r_2)^6$, $(r_5/r_4)^6$, $(r_4/r_2)^6$ ratios are evaluated by the data on imidazole geometry and values of Ni^{2+} ionic radius and Van-der-Waals radius of nitrogen atom.

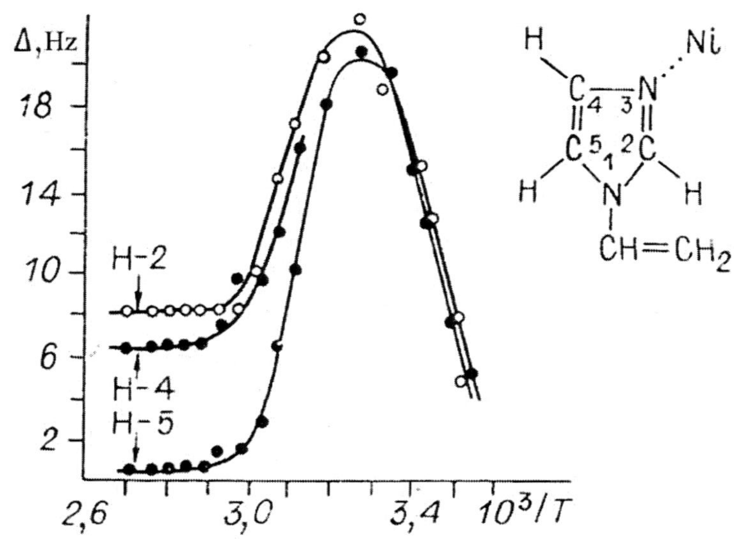

Figure 2.5. Temperature dependence of broadenings of the imidazole proton signals in the PMR spectra recorded with additive of paramagnetic complex $NiCl_2{\cdot}4VIM$.

Table 2.2.Temperature dependence of 1-vinylimidazole proton signals

T, K	δ', Hz			T, K	δ', Hz		
	H-2	H-4	H-5		H-2	H-4	H-5
283	1	-	1	330	23	26	24
286	1	-	1	336	22	26	21
290	1	-	1	342	22	26	21
293	4	-	2	348	22	25	21
300	10	-	14	352	21	25	20
306	14	-	19	357	21	25	19
314	23	-	27	363	19	23	18
322	24	27	27	370	18	22	16
325	23	27	26				

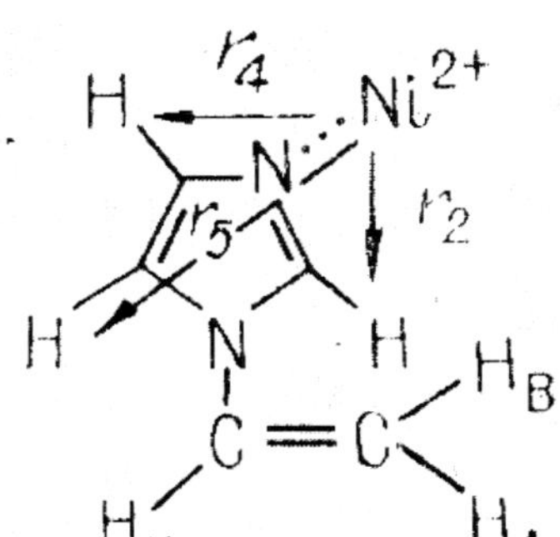

It has been found that $(r_5/r_2)^6 = 24$, $(r_5/r_4)^6 = 14$, $(r_4/r_2)^6 < 2$. It means that if the VIM molecule coordinates with $NiCl_2$ across $\boldsymbol{N-3}$ and the $\boldsymbol{\Delta}$ value of resonance lines is defined by the first member of the expression (2.14), Δ_2 and $\Delta_4 >> \Delta_5$ and $\Delta_2 > \Delta_4$.

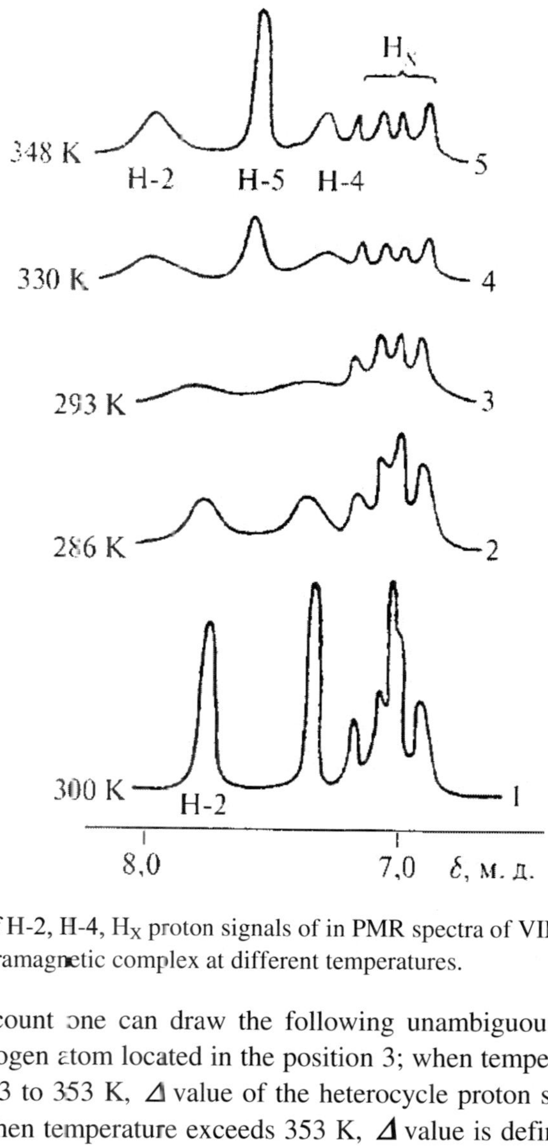

Figure 2.6. The positions of H-2, H-4, H$_X$ proton signals of in PMR spectra of VIM without (1) and with additives (2 – 5) of paramagnetic complex at different temperatures.

Taking this into account one can draw the following unambiguous conclusions: VIM coordinates with the nitrogen atom located in the position 3; when temperature of the studied sample changes from 313 to 353 K, Δ value of the heterocycle proton signals is defined by the expression (2.21); when temperature exceeds 353 K, Δ value is defined by first member of the expression (2.14). Since H-2 proton resonates in most low-field region, in the spectra of VIM, the signals of ring protons are observed as follows $\delta_2 > \delta_5 > \delta_4$.

Upon changing the temperature of the studied sample (with additive of the complex), signals of the vinyl group protons are not almost shifted relative to their positions in the spectrum of diamagnetic compound. Since these protons are considerably remote from the coordination ion, their resonance lines do not practically broaden, i.e., in all temperature ranges, the broadening of their signals is defined by the first member of the expressions (2.14).

The positions of H-2, H-4, and H-5 proton signals as well as HX proton of the vinyl group in the PMR spectra of VIM with and without additive of the paramagnetic complex of nickel at different temperatures is shown in Figure 2.6. In a similar mode there have been studied the changes in the PMR spectra of EIM, recorded with additives of the paramagnetic complex $NiCl_2 \cdot 4\,EIM$ at different temperatures. Also, the positions of the ring proton signals are shown in Figure 2.7.

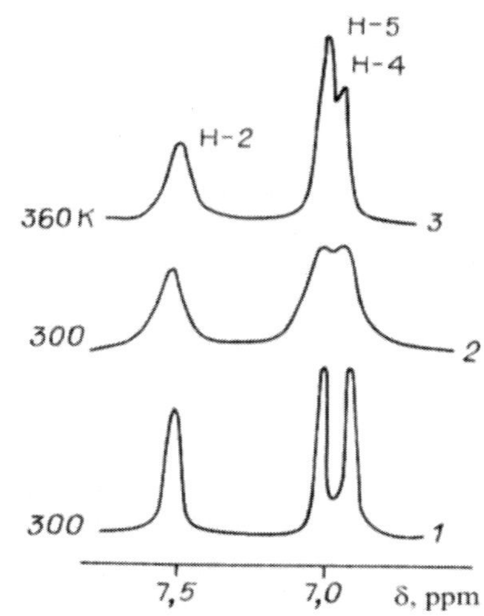

Figure 2.7. Fragment of the PMR spectrum of EIM without (1) and with additives (2,3) of paramagnetic complex at different temperatures.

The following two examples characterize the results obtained in the work [95]. Figure2.8 depicts the spectrum of 1,3,5-trimethylpyrazole recorded without and with the additive $NiBr_2 \cdot 3H_2O$ salt, which forms a paramagnetic complex with this pyrazole.

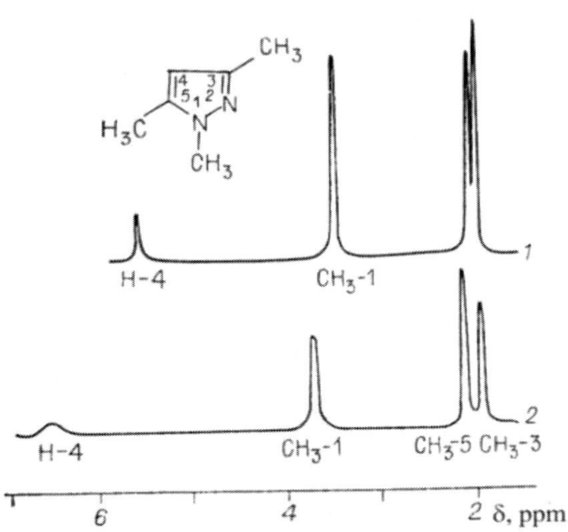

Figure 2.8. PMR spectrum of 1,3,5-trimethylpyrazole without (1) and with (2) additive of $NiBr_2 \cdot 3H_2O$.

As is seen from Figure 2.7 (curve 1), assignments of the 5-CH_3and3-CH_3 group signals is not unambiguous, since their protons resonating in the high field cause the appearance of two signals with insignificant difference of the chemical shifts (0,04 ppm). The complex formation leads to shift of these signals to the opposite sides and to different broadening.

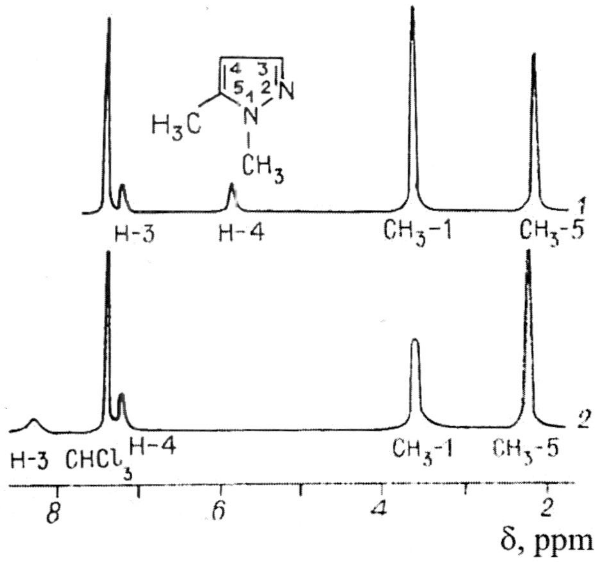

Figure 2.9. PMR spectrum of 1,5-dimethylpyrazole without (1) and with (2) additive of $NiBr_2 \cdot 3H_2O$.

From comparison of Δ and δ' values with the distance from coordination ion of resonating nuclei follows that a narrower signal belongs to methyl group protons in the position 5. Therefore, in the spectrum of 1,3,5-trimethylpyrazole recorded without the paramagnetic complex additive, a singlet in the high-field is caused by protons of the 3-CH_3 moiety.

Figure 2.9 shows the spectra of an isomer of dimethylpyrazole recorded without and with the additive of the above-mentioned nickel salt. Since the broadening of H_4proton signal, which position in the spectrum is known, is significantly lesser than that of a signal of other pyrazole proton, this isomer represent 1,5-dimethylpyrazole, where $r_3^6 << r_5^6$.

It should be noted that the method for interpretation of the NMR spectra of complex organic compounds described here is based on the application of peculiarities of the NMR spectra of paramagnetic complexes, and hence is quite reliable and simple. Before publication of the work [95], there were no reliable methods for identification of pyrazole isomers formed by hydrogen substitutions in the positions 5 and 6. These isomers have been identified for the first time in 1926. However, in 1956 it was suggested that the previous identification of 1,3- and 1,5-dimethylpyrazoles was incorrect. Since 1963, the NMR method is applied for structural studies of pyrazoles. The comparison of chemical shift values and spin-spin coupling constants in the spectra of substituted pyrazoles has confirmed that the initial identification of the aforementioned isomers is correct. But this confirmation has indirect character.

The problem of 1,3- and 1,5-dimethylpyrazole identification is solved unambiguously using the method for determination of functional group positions in the molecules of complex organic compounds capable of forming the paramagnetic complexes. Moreover, interpretation of 5-ethynyl-1-methylpyrazole NMR spectrum with the help of this method allows assigning the resonance lines in the spectra of disubstituted pyrazoles, which do not form paramagnetic complexes. The method of paramagnetic additives makes it possible to reveal specifics of the vinylation reaction involving pyrazole and indazole as well as to establish the peculiarities of the PMR spectra of the reaction products [172].

Of special interest is the following feature of the PMR spectra of 1-vinylpyrazole (VIP), recorded with small amounts of nickel acetylacetonate $Ni(AA)_2$ additive [173]. It turns out that when concentration of the paramagnetic complex in sample increases, the doublets from $\beta-$olefinic protons are replaced by singlets, which line widths are lower than the values of vicinal constants $J = 9{,}0$ Hz, $J = 15{,}8$ Hz (Figure 2.10). The explanation of this effect is based on the idea on alteration of relaxation time T_2 of one of interacting (according to spin-spin mechanism) nuclei in the samples containing paramagnetic ions [174]. Such alteration of T_2 is owing to chemical molecular exchange between the coordination sphere of paramagnetic complex and free ligand. The condition necessary for removal (suppression) of spin-spin coupling is expressed as follows:

$$JT_{2P} < 1, \tag{2.26}$$

where $1/T_{2P} = 1/T_{2M} - 1/T_2$; T_{2M} and T_2 are transverse relaxation times of the nucleus on and out of the paramagnetic complex.

Under the conditions of fast exchange and short times of electron spin relaxation we have the following expression: $1/T_{2P} = 1/T_{2M}$ (see section 2.1 of this monograph). Therefore, the expression (2.26) can be presented as shown below:

$$J < P_M / T_{2M} \, . \tag{2.27}$$

According to the expression (2.15), in the absence of contact contribution, relaxation time T_{2M} is directly proportional to r^6. If in the ligand molecule, the nucleus i, interacting with the nucleus j, is located quite close to the coordination ion, the inequality (2.27) can be realized. As a consequence, the constant J_{ij}, which is expressed in units of frequency, will not be observed in the signal of the nucleus j. Here it is pertinent to remind that T_{2M} is the relaxation time of resonating nucleus located in the first coordination sphere of the paramagnetic complex. The increase in this time will lead to the situation when the effect of the magnetic moment of unpaired electron on the nucleus becomes more prolonged. Eventually, this nucleus will be transferred between the permitted states significantly quicker with all the consequences following from this. The location of unpaired electron spin density on the nucleus i can strengthen the effect of spin-spin coupling suppression. Certainly, this will be legalized, provided that the signal from j nucleus is not significantly broaden for some other reason, for example, due to delocalization of unpaired electron spin density on this nucleus.

After the addition of small amount of $Ni(AA)_2$ to the VIP solution, the signal of H_X proton strongly broadens (in Figure 2.10, curve 5, it is almost visible) that testifies to the essential change of T_2 value for H_X. It is also indicative of such disposition of the vinyl group and heterocycle, when H_X is oriented towards the N-2 atom. Just with the latter VIP coordinates with $Ni(AA)_2$.

Addition of nickel to a solution of acetylacetonate leads also to the suppression of spin-spin coupling between H-3 and H-4 protons, therefore, a signal of H-proton turns from triplet into doublet (*cf.* spectra in Figure 2.10, curves 1 and 3).

The method of paramagnetic additives described here can be also applied to other heteroatom connections. It is natural that for different ligands and coordination ions which can form the complexes, which differ by ω and T_{2M} values, the question of approximation of the expressions (2.2) and (2.3) with formulas (2.16) – (2.22) should be solved differently for each specific case. Since the complexes formed by elements of the first transition group have essentially different times of electron relaxation (τ_s (τ_s^{Mn} и $\tau_s^{Cu} >> \tau_s^{Co}, \tau_s^{Ni}, \tau_s^{Fe}$) [1,5]), comparison of Δ and δ' values for the purpose of resonance lines assignment can be made at different temperatures of the studied samples. This is shown in Figure 2.11 [160].

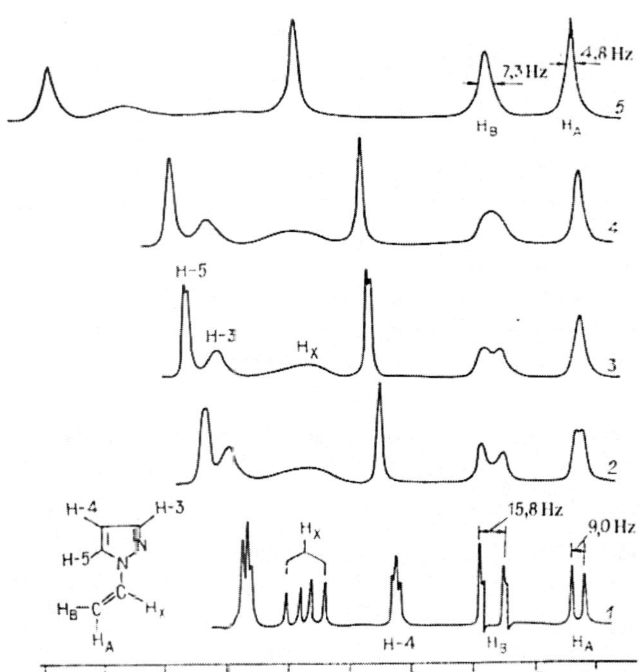

Figure 2.10. PMR spectra of 1-vinylpyrazole without (1) and with (2-5) additives of $Ni(AA)_2$.

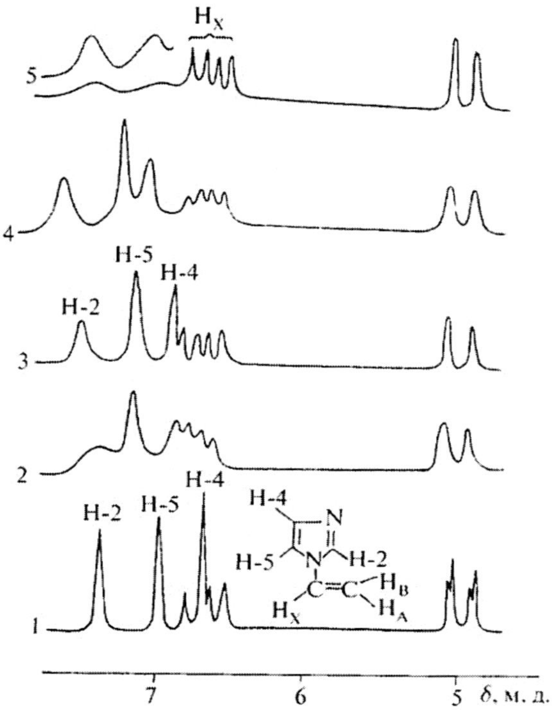

Figure 2.11. PMR spectra of VIP (signal of H_A proton is overlapped with signal of the solvent, nitromethane) without (1) and with additives of paramagnetic complexes of nickel (2), cobalt (3), iron (4), copper (5).

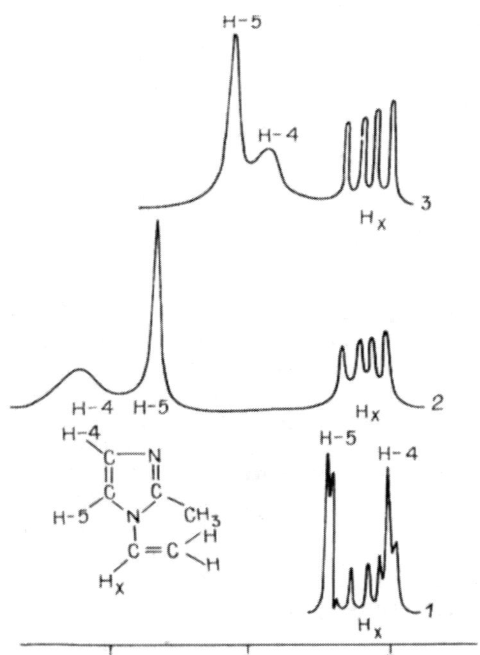

Figure 2.12. A fragment of PMR spectrum of 1-vinyl-methylimidazole without (1) and with additive of paramagnetic complex of cobalt (2) and nickel (3) recorded at room temperature.

The similar situation can take place for the same coordination ion, but with different compounds acting as ligands. For example, in the PMR spectra of pyrazole derivatives, the values of Δ и δ' at room temperature correspond to fast exchange that is illustrated by Figure 2.12 [161]. However, in the PMR spectra of VIM and EIM, recorded with additives of the paramagnetic complex, these conditions are realized only are a temperature exceeding 70° C.

2.2.1. 1-Methyl-2-Aminobenzimidazole

^{13}C and ^{1}H spectra of 1-methyl-2-aminobenzylimidazole (XIV) recorded in the presence of paramagnetic complex $\boldsymbol{CoCl_2 \cdot 4\,EIM}$ (XV) have been studied [118].

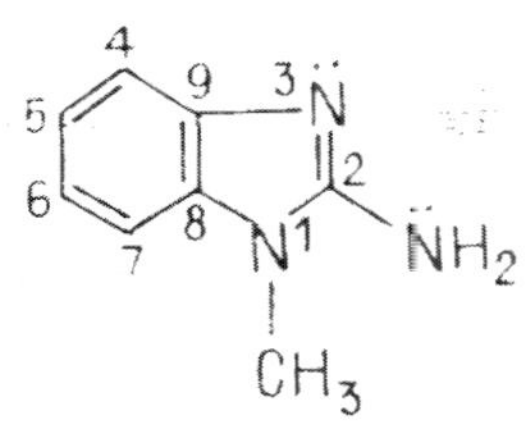

XIV

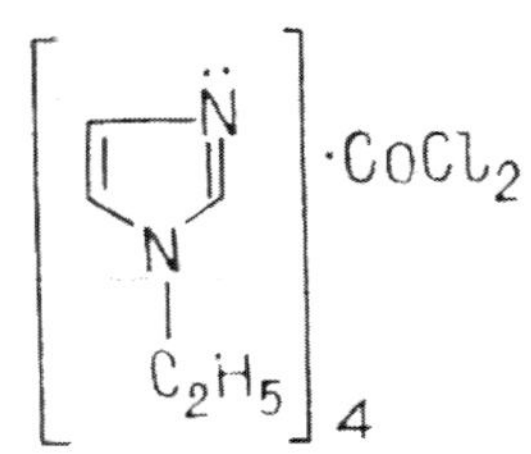

XV

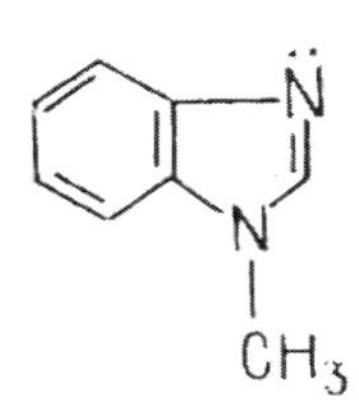

XVa

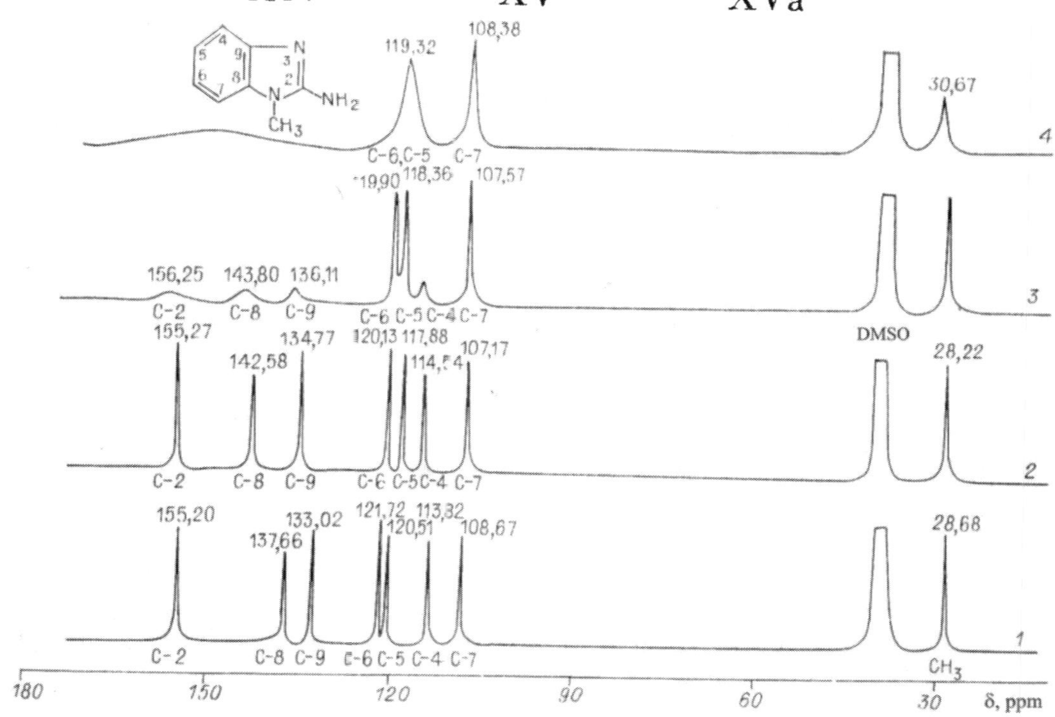

Figure 2.13. ^{13}C NMR spectrum of $ZnCl_2 \cdot 2$(1-methyl-2-aminobenzimidazole) complex (1) and ^{13}C NMR spectra of 1-methyl-2-aminobenzimidazole without (2) and with different additives of paramagnetic complex (3,4).

It turns out that the addition of paramagnetic complex XV leads to low-field shift of all signals in the ^{13}C– {^{1}H} NMR spectrum of compound XIV (except for a signal from C-6 carbon atom). Abnormal shift and broadening of lines from C-4 atom as compared to lines of C-5, C-6 and C-7 atoms (figure 2.13) is observed. Such character of the spectra transformation indicates the delocalization of unpaired electron spin from the σ – system of the benzimidazole cycle owing to coordination of N-3 nitrogen atoms [18]. At the same time, a shift of C-6 signal toward high-field testifies to the participation of π – system in HFC transfer across the molecule. This is also supported by the PMR spectrum of compound XIV, which transformation with paramagnetic additives XV leads to both high-field and low-field shift of the proton signals (Figure 2.14).

A possibility of spin density localization in the π – system due to effects of polarization has been stressed in the work [35]. However, π – mechanism of the delocalization can take place owing to the interaction of molecule XIV with paramagnetic ion of nitrogen atom of the amino group [175]. In this connection, the PMR spectra of 1-methylbenzimidazole XVa transformed by the complex XV have been analyzed. It is found that nature of paramagnetic shifts in the spectrum of this compound, which should be coordinated with compound XV via pyridinic N-3 nitrogen, is similar to that in the spectrum of molecule of XIV (Figure 2.15).

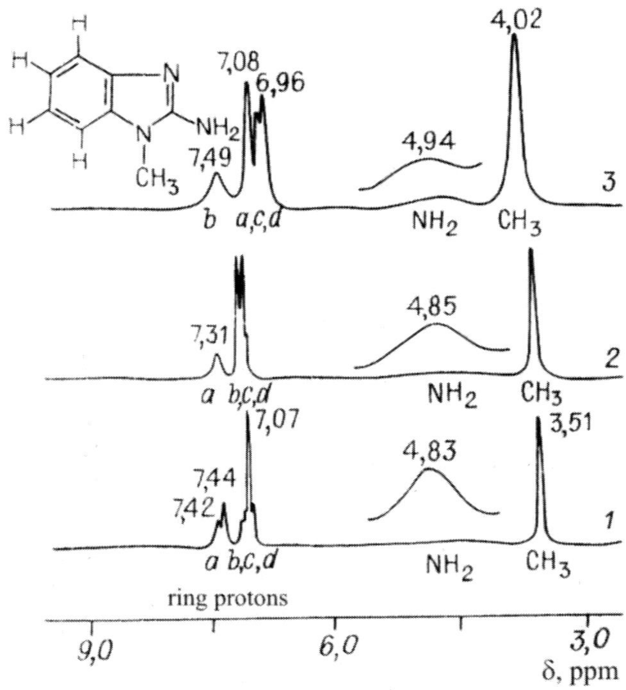

Figure 2.14. PMR spectra of 1-methyl-2-aminobenzimidazole without (1) and with different additives of paramagnetic complex (2,3).

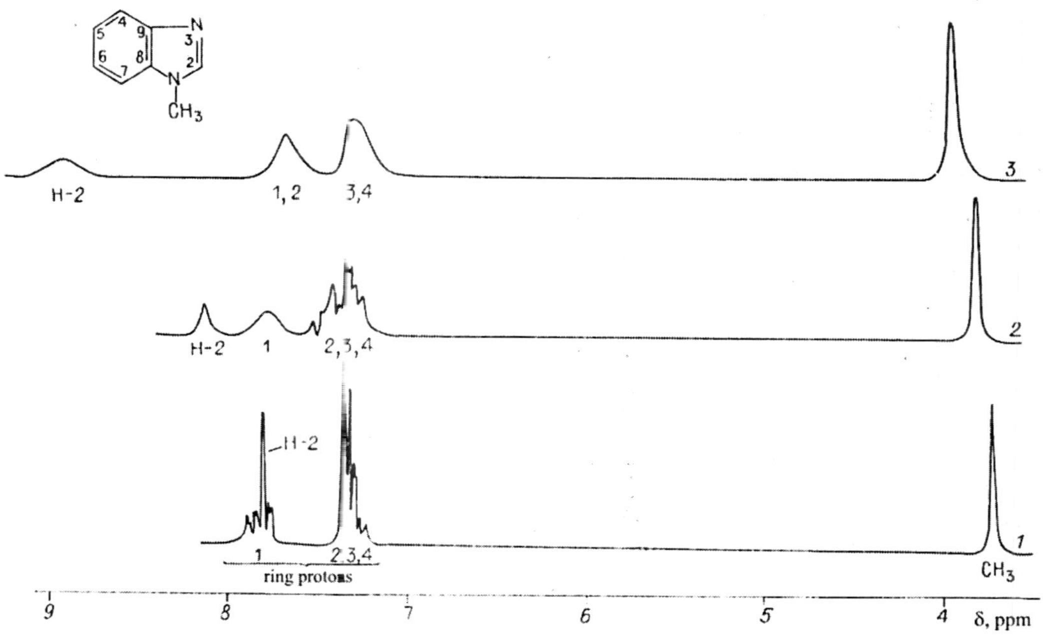

Figure 2.15. PMR spectra of 1-methylbeenzimidazole without (1) and with different additives of paramagnetic complex (2,3).

Thus, localization of the uncompensated electron spin in the $\pi-$system of ligand XIV is apparently explained by peculiarities of electronic structure of complex XV and specifics of HFC in it, but not by coordination across the nitrogen atom of the amino group. Such conclusion is confirmed also by the fact that protons of the amino group in molecule XIV undergo insignificant low-field shift, whereas formation of the coordination bonding aniline··· $\mathbf{Co}^{2+}$ is accompanied by the high-field shift of the NH_2-group proton signals [175].

In addition to the investigations of NMR spectra (transformed by paramagnetic additives) of compound XIV, in work [118], the comparison of the ^{13}NMR spectrum of molecule XIV with $ZnCl_2$ has been performed (see Figure 2.13, curve 1). Earlier, the study of organylhalostannane complexes with VIM and EIM has revealed that the coordination at N-3 atom leads to the absence of C-2 atom signal, whereas the C-4 resonance is high-field shifted [176]. In the studied diamagnetic zinc complex, a similar picture is observed. The latter fact can be considered as another proof in favor of localization of the coordination bond on N-3 atom of molecule XIV.

The data obtained allow to conclude that in non aqueous solutions ($CHCl_3, DMSO$), 2-amino-1-methylbenzimidazole with two-valence cobalt and zinc chlorides forms complex compounds, in which the coordination bond is localized on endocyclic nitrogen atom of the pyridine type. Taking into account the results of previous investigations into synthesis and structural features of 2-amino-N-alkyl(aralkyl)benzimidazole complexes with metal halogenides of III-V groups in crystalline phase, the method for localization of coordination bonding discussed here should be considered as a general one, which does not depend upon the metal nature and aggregate state of the medium where the complex formation of 2-aminobenzimidazoles [118] is studied.

2.2.2. 1,1-Divinyl-2,2-Biimidazolyl

In the work [177], the complex forming additives are applied to elucidate the denticity issue related to 1,1-divinyl-2,2-biimidazolyl XVI as well as to determine spatial structure of the same compound and its complexes with transition metal salts.

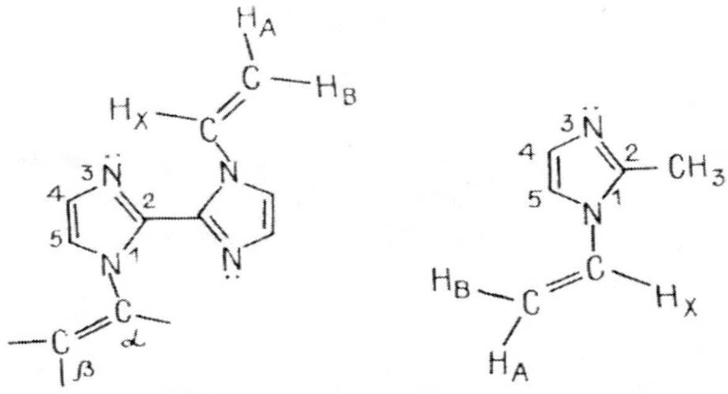

In the PMR spectrum of 1,1-divinyl-2,2-biimidazolyl complex XVII, the descreening effect of the N-3 atom lone pair on H_X nucleus (0,65 ppm) remains. As shown in the above scheme, the same effect should also retain in the case of trans-orientation of molecule XVI cycles and its vinyl groups. It should be noted that the descreening of H_X nucleus is not observed upon comparison of this proton signal shifts in the PMR spectra of divinylbenzimidazole hydrochlorides and VIM, where lone pairs of N-3 atoms participate in the formation of donor-acceptor bonding. The decrease in temperature of the sample containing a solution of complex XVII to 0° C leads to strong broadening of the H_X signal. At this temperature, the fast exchange with molecules XVI between the coordination sphere and solution becomes impossible. As a consequence, nonequivalence of the H_X begins to be observed in the spectrum. This phenomenon should manifest itself in the case of molecule coordinated in monodentate manner. The data testifying to the monodentate character of ligand XVI have also been obtained by the method of paramagnetic additives using a manganese complex.

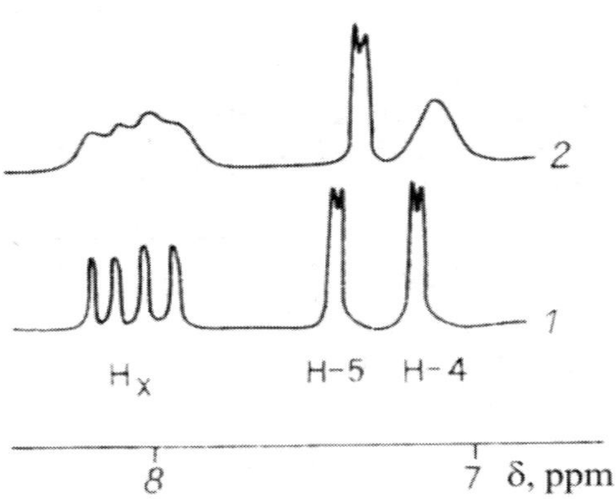

Figure 2.16. A fragment of PMR spectra of compound XVI without (1) and with (2) additive of $Mn(AA)_2$.

The addition of these additives to a solution of 1-vinyl-2-methylimidazole XVIa leads to incomparably stronger broadening of the H-4 signal in comparison with the H-5 signal, since H-4 nucleus is located much closer to the coordination center (N-3 nitrogen) than H_X nucleus (here it is pertinent to note that the broadening is proportional to r^{-6}, $r-$ is a distance from the coordination ion to resonating nucleus). However the same paramagnetic additives cause one order broadening of H-4 and H_X proton signals in the spectra of compound XVI (Figure 2.16). Therefore, in compound XVI, the H_X and H-4 nuclei are remote from the paramagnetic center by the same distance (i.e. $r_4 \approx r_X$) that indicates the monodentate character of the ligand disposition in the coordination sphere of Mn^{2+}.

2.3. Lanthanide Paramagnetic Additives

In principle, the application of paramagnetic complexes of rare-earth elements is based on the same theoretical ideas that have been formulated for the complexes of iron group. Unlike the latter, lanthanides used for the analysis of NMR spectra of organic heteroatom compounds have very short electron relaxation times (except for Gd^{3+} and Sm^{3+}) [178] that significantly facilitates the task of recording the spectra with considerable values of paramagnetic shifts. In this regard, paramagnetic complexes of rare-earth elements or LSR possess indisputable advantage [7]. The results of the LSR method application for the solution of structural tasks are discussed below. The discussion is preceded by three examples taken from practice of the monograph authors, which illustrate how the usage of europium tris(diketonate) $Eu(DPM)_3$ increases informational content of the NMR spectra.

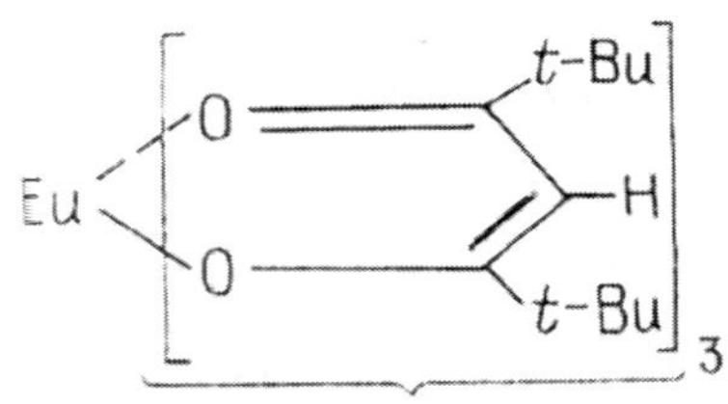

DPM

2.3.1. 1-Ethyl-2-Methylimidazole

Figure 2.17 shows the PMR spectra of 1-ethyl-2-methylimidazole. Its interpretation is not difficult. In the spectra, triplet ($\delta = 1{,}38$ ppm) and quartet ($\delta = 3{,}85$ ppm) appear due to protons of the methyl and methylene moieties of the ethyl radical, respectively. The singlet in the high-field ($\delta = 2{,}28$ ppm) is assigned to the heterocycle methyl protons. A characteristic feature of the spectrum is coincidence of signals position for H-4 and H-5protons that leads to appearance of singlet in low field ($\delta = 6{,}68$ ppm). As a consequence, a question about of character of the anticipated spin-spin coupling between protons H-4 and H-5 becomes obscure. The problem can be solved if to record the spectrum of the sample containing $Eu(DPM)_3$ additive. The matter is that H-4 and H-5 protons are remote from N-3 atom (with which the molecule coordinates with Eu^{3+}) by different distances. Eventually, this leads to chemical nonequivalence of these protons in the complex and different signal shifts relative to position in the spectrum of sample recorded without the additive. As a result, the above protons give separate signals allowing spin-spin coupling constant to be easily determined (J_{45} = of 1,3 Hz, Figure 2.17, spectrum 2).

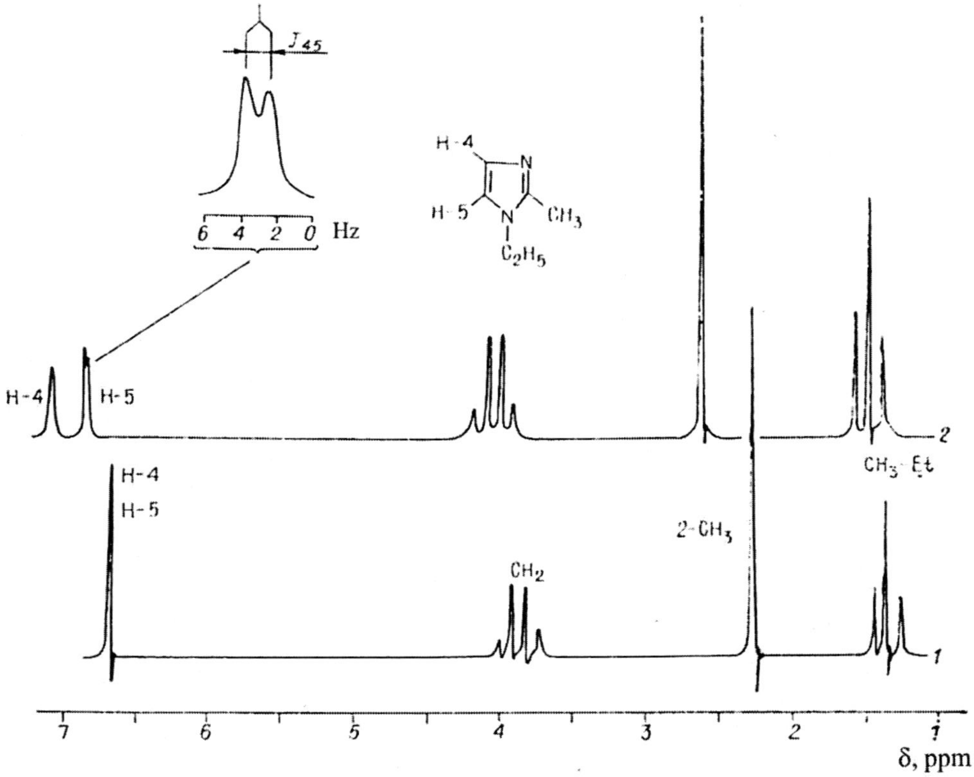

Figure 2.17. PMR spectra of 1-ethyl-2-methylimidazole without (1) and with (2) additive of $Eu(DPM)_3$.

2.3.2. N,S-Divinil-2-Mercaptobenzimidazole

The PMR spectrum of N,S-divinyl-2-mercaptobenzimidazolerecorded without the paramagnetic additive is shown in Figure 2.18, curve 1. Since the vinyl group protons at nitrogen and sulfur atoms give characteristic signals (according to position and multiplicity), their assignment using the double resonance technique does not present difficulties. Signals of the benzene ring protons overlaps with each other as well as with $H_{X'}$ and H_X proton signals give not unresolved multiplet in the region 6,87 – 7,65 ppm. The addition of $Eu(DPM)_3$ to the sample leads to noticeable shift of the signal positions relative to the spectrum depicted in Figure 2.18, curve 1. Signals of the ring protons are not overlapped with each other, therefore, they can be identified (Figure 2.18, curve 2).

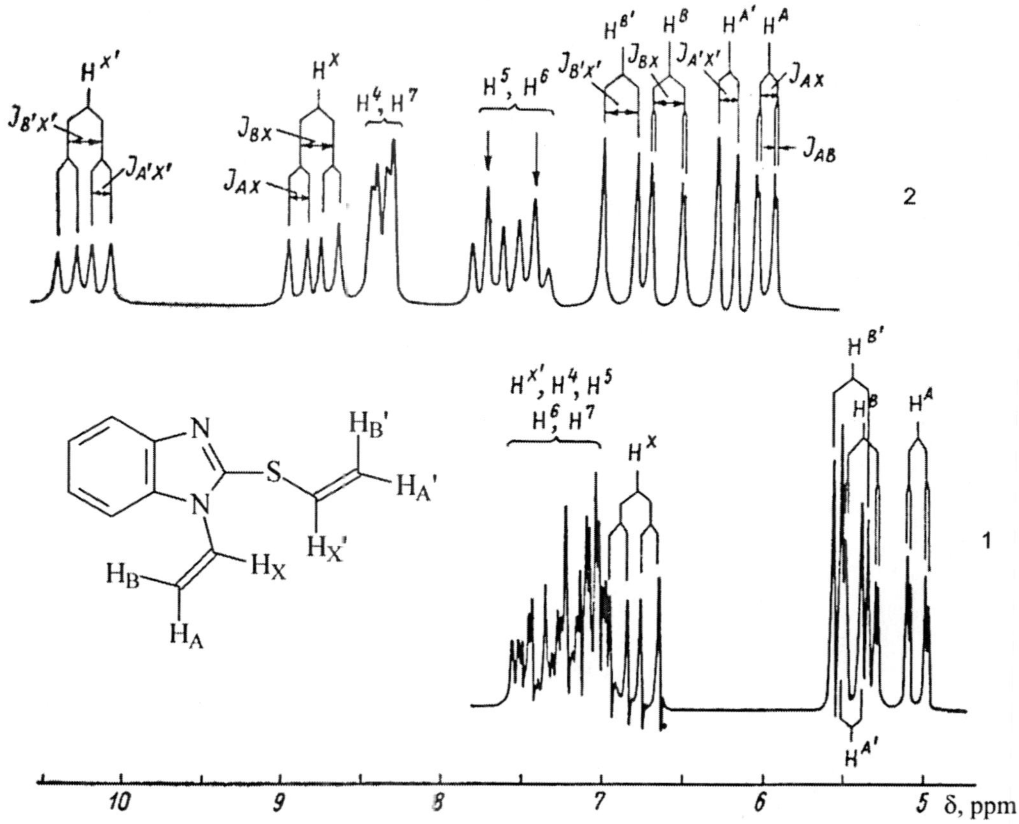

Figure 2.18. PMR spectrum of N,S-divinylmercaptobenzimidazole recorded without (1) and with (2) $Eu(DPM)_3$ additive.

2.3.3. Vinyl Derivatives of Imidazole, Pyrazole, Triazole

It has been established [179] that in the PMR spectra of 1-vinyl derivatives of imidazole and pyrazole recorded at a temperature above 0° C, the broadenings of H_A proton and ring proton signals are observed (H-2 in the spectrum of VIM, H-3 and H-4 in the spectrum of VPZ). These broadenings are not suppressed by additional radiation (homonuclear double resonance), but depends on temperature of the studied sample.

For example, in the spectrum of VIM recorded at – 60° C, line width of quartet from H_A and H_B is identical, and form of the H-2 signal is defined by spin-spin coupling between protons of the heterocycle (Figure 2.19, 2.20).

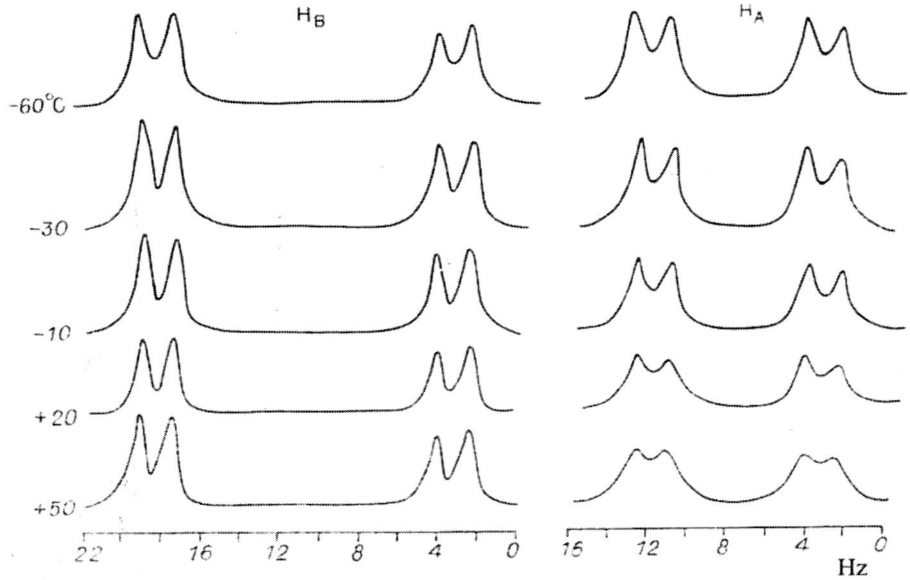

Figure 2.19. Signals of the vinyl group *β* – protons in the PMR spectrum of VIM recorded at different temperatures of the studies sample.

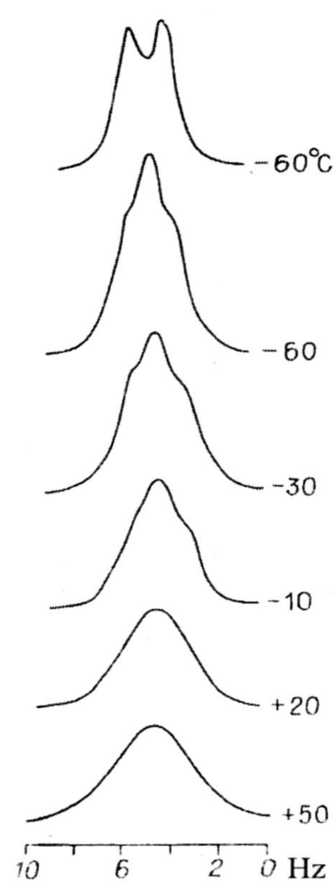

Figure 2.20. Temperature dependence of H-2 proton signal width in VIM; at – 60° C H-2 signal is shown as decoupled from H-5 proton.

In the spectrum of VPZ, recorded at the same temperature, the above mentioned broadening is not suppressed (figure 2.21). However, application of the double resonance technique turns out to be effective since it cause narrowing of H_A quartet lines (Figure 2.21, curve a). This fact is indicative of the long-ranged spin-spin coupling between protons of the ring and vinyl group. Effect of temperature on width of the resonance lines testifies to influence of the quadruple moment of the nucleus on proton relaxation of through spin-spin coupling communication $^1H-{}^{14}N$.

Indeed, in the $^1H-\{^{14}N\}$ spectrum of VIM, the additional broadening of H_A (Figure 2.22) and H-2 signals is absent. From the experiment follows that H-2 and H_A interact with the same nitrogen atom. To elucidate with which nitrogen atom interact these protons, the PMR spectrum of VIM has been recorded with additive of $Eu(DPM)_3$. It appears that signals of H-2 and H-4 protons (closest to the coordination center, N-3 nitrogen) are shifted by 1,71 and 1,65 ppm, respectively, while signal of the nitrogen interacting with H-2 and H-4 is shifted by 8,9 ppm. In paramagnetic complex of pyridine with $Eu(DPM)_3$, ω_N shift value of ^{14}N signal (pyridine coordinates with the central ion via this atom) is 1500 ppm [180], and $\omega_2=\omega_4=12{,}16$ ppm[181]. Therefore, shift of H-2 and H-4 by 1,5 ppm in the PMR spectrum of VIM should correspond to approximately 150 ppm shift of ^{14}N. In this case, ω_N value is much lower than the expected values. This testifies to the spin-spin coupling of H_A and H-2 with N-1 atom nucleus.

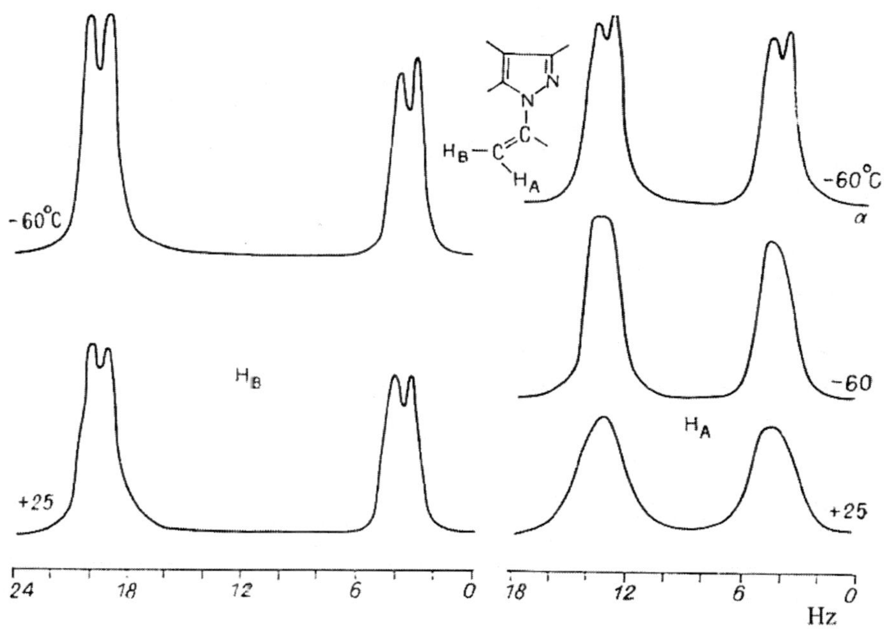

Figure 2.21. A fragment of vinylpyrazole PMR spectrum recorded at different temperatures (a). Signal of H_A proton decoupled from H-3 proton.

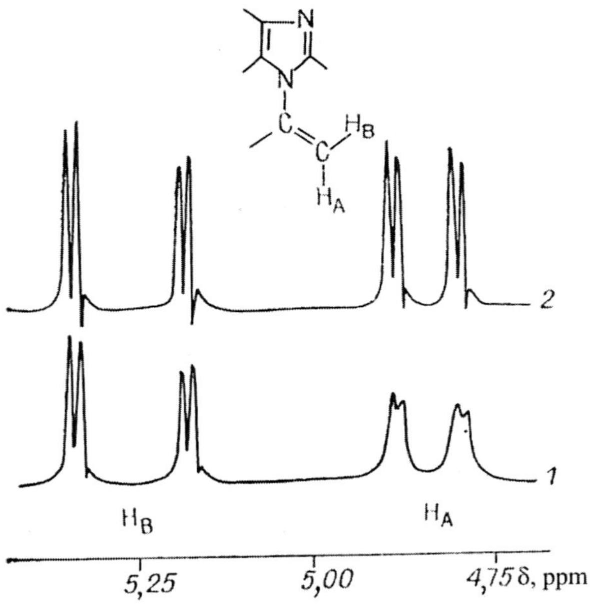

Figure 2.22. Shape of H_A and H_B proton signals in normal PMR spectrum of VIM(1) and in the $^1H - \{^{14}N\}$spectrum (2).

In the PMR spectrum of VPZ recorded with $Eu(DPM)_3$ additive, when ω_3 = 0,48 ppm, the signal of nitrogen nucleus, assigned [179] to N-1 atom and interacting with H_A , is shifted by 8,3 ppmin the direction opposite to the shift of pyridine ^{14}N. From the VPZ$^1H - \{^{14}N\}$ spectrum that H-3 couples both with N-1 and N-2, the signal of which is by 93 ppm signal low-field shifted relative to the position of the resonance line N-1. When $Eu(DPM)_3$ is added to the sample, this shift is 22,2 ppm shifted towards high-field (like in pyridine) and ω_3 =0,48 ppm. This value is approximately 3 times lower than the expected one if to be guided by a shift of ^{14}N signal in the spectrum of pyridine. This can be due to the effect of substituent in $\alpha-$position relative to the coordination center. It should be noted that proton H-5 also couples with nitrogen in the position 2. Thus, using the techniques of heteronuclear double resonance and paramagnetic additives it has been established that in the PMR spectrum of VIM, the additional broadening of signals is caused by the spin-spin coupling between ^1H and ^{14}N, while in the spectrum of VPZ, the broadening can be also rationalized by the long-range spin-spin coupling $^1H - {}^1H$ via six bonds. Here it is pertinent to emphasize that the analysis of ^{15}N NMR spectra has revealed [182] that in VIM and 1-vinyl-2-formylimidazole, there is the spin-spin coupling between the vinyl group protons and N-1 atom (J_{NH} =5 Hz).

In vinyl derivatives of triazole, the quadruple relaxation of ^{14}N nuclei also causes additional broadening of signals of the vinyl group and heterocycle protons, which changes upon varying a temperature of the sample. Such change is observed upon recording the PMR spectra under the conditions of decoupling from nitrogen, i.e. in the $^1H - \{^{14}N\}$ spectra (Figure 2.23, 2.24). In this case, similar to VIM, long-range spin-spin coupling between H_A and the heterocycle proton is not observed. However in the spectrum of 1-vinyltriazole, long-range spin-spin coupling between $\alpha-$proton (H_X) of the vinyl group with one of the ring protons is realized. As a consequence, its signal as well as quartet from H_X becomes broaden in the $^1H - \{^{14}N\}$ spectrum (Figure 2.24, curve 2). At low temperatures (– 20° C and below), lines of a quartet have doublet structure which disappears under the conditions of homonuclear decoupling from the heterocycle proton in the position 5 (Figure 2.25, curve 2, $\boldsymbol{J}_{5X}$ = 0,6 Hz). In this connection, 1-vinyl-2-methylimidazole is of special interest. In the PMR spectra of this compound, H_A proton signal together with quartet lines from H_X proton are broaden (Figure 2.26). Such broadening (unlike H_A), like in 1-vinyltriazole, is not suppressed even at decrease of the sample temperature. From the double resonance experiments and values of paramagnetic shifts follows that the broadening is caused by long-range spin-spin coupling of HX with H-5 via four bonds ($\boldsymbol{J}_{5X}$ = 0,6 Hz) and with H-4 via five bonds ($\boldsymbol{J}_{4X}$ =0,75 Hz).

Thus, the application of paramagnetic additives allows not only to assign proton signals of the studied compounds (it is a non-trivial task), but also to establish their characteristic feature, namely their ability to show long-range spin-spin coupling through five and six bonds in the PMR spectra. Character of heteronuclear spin-spin coupling $^1H - {}^{14}N$ has also been determined.

The coupling between protons of the heterocycle and vinyl group is observed in the PMR spectrum only in the case, if nitrogen atom (pyrazole, triazole) is located in the position 2 of the five-membered cycle, or if the methyl group (imidazole) is substituent at the C-2 atom. It should be stressed that the effect of quadruple relaxation of ^{14}N nucleus is also manifested itself in the spectrum of EIM. Indeed, at 40° C, the H-2 and H-4 protons cause the appearance of broaden signals, whereas at – 100° C signals of these protons have the resolved triplet structure.

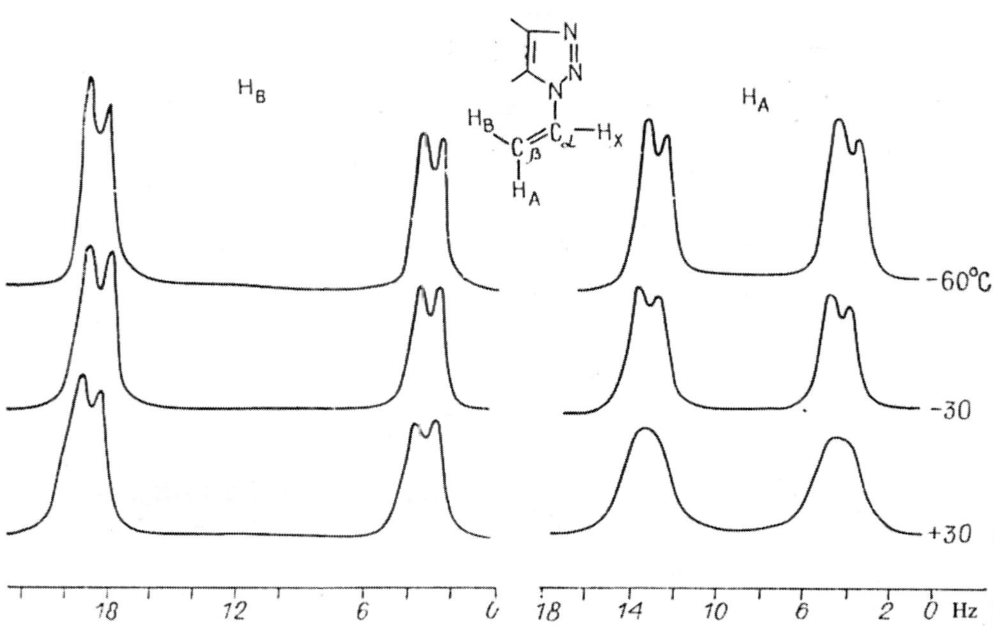

Figure 2.23. Signals of vinyl group β – protons in the PMR spectrum 0f 1-vinyltriazole recorded at different temperatures of the studied sample.

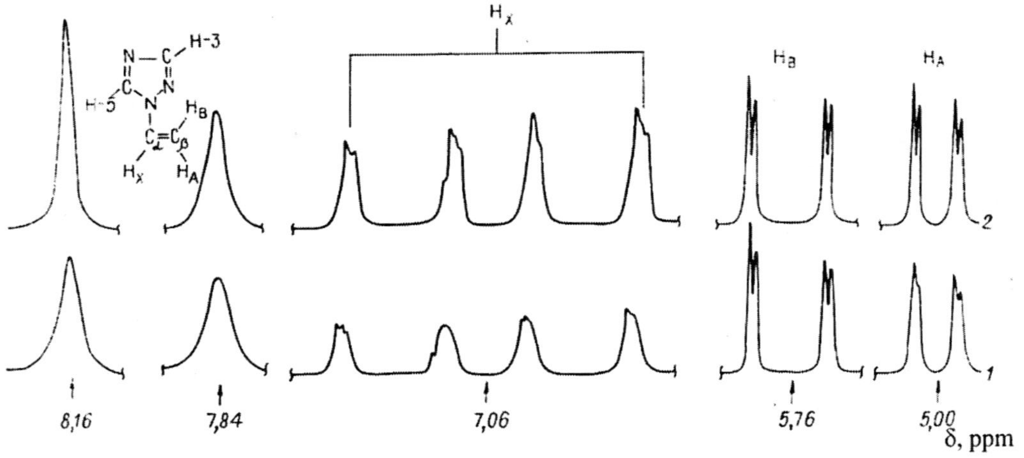

Figure 2.24. Normal PMR spectrum of 1-vinyltriazole (1) and the spectrum recorded with decoupling from ^{14}N nucleus (2).

The employment of complex forming additives method can lead eventually to obtaining such spectral information, which further (without application of the paramagnetic additives method) can be considered as a tool for structural studies. This is a case of vinyl derivatives of azoles, in particular, 1-vinylpyrazole, the NMR spectra of which are analyzed and discussed below. It should be noted that S-*cis-trans*-isomerism is inherent in the compounds, in which the heteroaromatic nucleus is conjugated with acyclic unsaturated fragment. Actually, similar information about 1-vinylpyrazole was absent before the publication of the work [183]. The only exception is the paper [184], where it has been reported that 1-methyl-2-nitro-5-vinylimidazole in a crystalline state exist as a mixture of two isomers. Besides, it is assumed [185] that 1-vinylpyrazoles should most likely exist in S-*cis*-configuration, while in the work [186] it is concluded on conformational homogeneity of the mentioned compounds and it is suggested that they can be formed exclusively as S-*trans*-isomers.

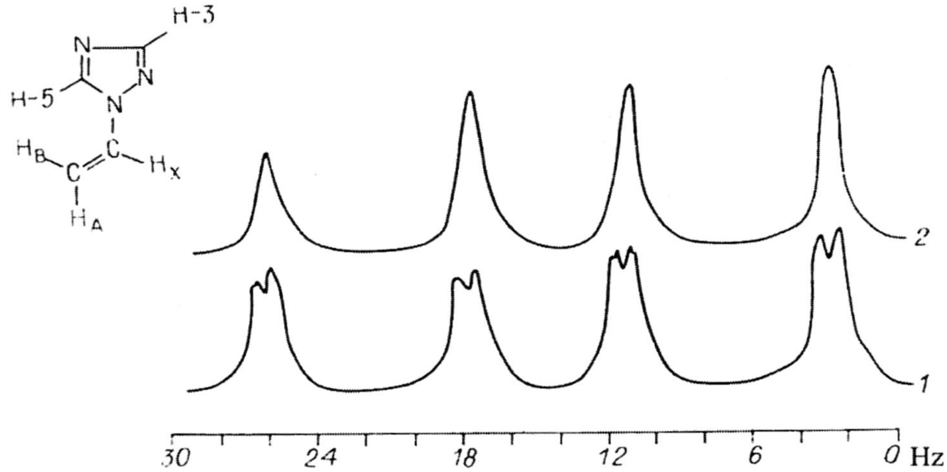

Figure 2.25. H_X proton signals of the vinyl group of 1-vinyltriazole without (1) and with (2) decoupling from H-5 proton at – 20^0 C.

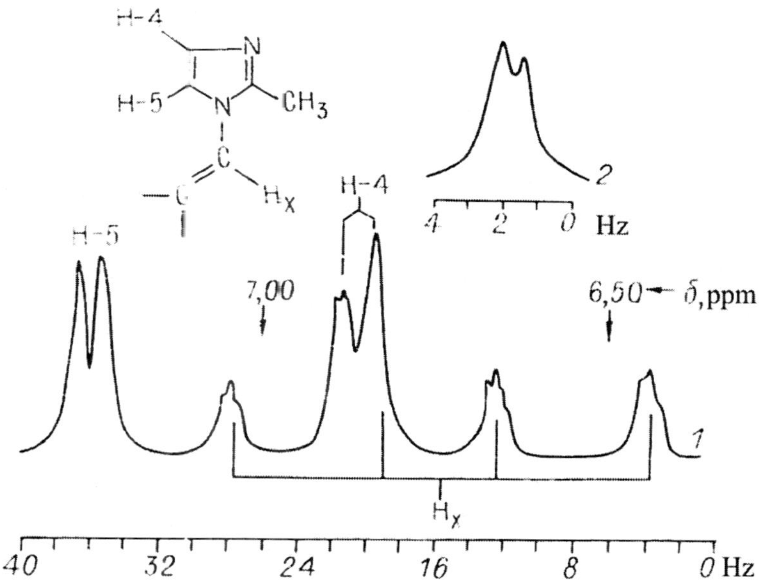

Figure 2.26. A fragment of the PMR spectrum of 1-vinyl-2-methylimidazole (1) and one of the lines of H_X quartet under conditions of deccupling from H-5 proton (2).

Table 2.3. Alterations of Δδ values in the PMR spectra of 1-vinylpyrazoles

Difference in shifts	Proton	Δδ, ppm.			Difference in shifts	Proton	Δδ, ppm		
		CCl_4	$CDCl_3$	CD_3OD			CCl_4	$CDCl_3$	CD_3OD
δ(XX) –	H-4	0,26	0,21	0,21	δ(XXIV) –	H-4	0,20	0,20	0,23
δ(XXVIII)	H-5	0,16	0,15	0,18	δ(XXIII)	H_X	0,11	0,08	0,12
	H_X	0,13	0,12	0,14		H_A	0,10	0,09	0,12
	H_A	0,10	0,10	0,12		H_B	0,08	0,10	0,11
	H_B	0,09	0,08	0,10		H-4	0,28	0,30	0,26
δ(XXIII) –	H-3	0,15	0,15	0,14	δ(XXIV) – δ(XX)	H_X	0,11	0,08	0,03
δ(XXVIII)	H-4	0,34	0,31	0,27		H_A	0,02	0,00	0,00
	H_X	0,13	0,12	0,05		H_B	-	- 0,15	- 0,02
	H_A	0,02	0,01	0,00		H_X	0,19	- 0,40	- 0,37
	H_B	- 0,18	- 0.17	- 0 03	δ(XXV) – δ(XXII)	H_A	-	- 0,33	- 0,35
δ(XIX) –	H-3	- 0,12	- 0.10	- 0.40		H_B	0,15	- 0,09	- 0,15
δ(XXVIII)	H-5	0,05	- 0,01	0,29			-		
	H_X	0,07	0,03	0,07			0,42		
	H_A	- 0,05	- 0,10	- 0,04			-		
	H_B	0,02	- 0,02	- 0,03			0,33		

Note: Deshielding is marked as « – ».

In this connection, the NMR study of spatial structure of 1-vinylpyrazole has been conducted. The study is based on results of application of the paramagnetic additives method (see section 2.2. of this monograph). It is found that J_{AB} constants, characterizing spin-spin coupling of $\beta-$olefinic protons in the 1-vinylpyrazole XVIII, 1-vinyl-4-bromopyrazole XIX, 3-methyl-1-vinylpyrazole XX and 2-vinylimidazole XXI, have close values, but considerably differ from values of J_{AB} constants of 5-methylsubstituted 1-vinylpyrazoles and 1-vinylimidazole XXII. Thus, the internal chemical shift $\delta'' = \delta_B - \delta_A$ in the spectra of compounds XVIII-XXI also coincides, but differs from this value in the spectra of 5-methyl-1-vinylpyrazole XXIII, 3,5-dimethyl -1-vinylpyrazole XXIV, 4-nitro-3,5-dimethyl-1-vinylpyrazole XXV and compound XXII. This allows assuming that relative orientation of the vinyl fragment and heterocycle in the above series of vinylpyrazoles is different.

Character of proton signal shifts of the vinyl group also changes and depends not only on nature of the substituent, but also on its position in the cycle (table 2.3). It is known that the protons, which are not directly coupled with nitrogen atom containing lone electron pair, but which are located near this atom, are significantly shifted towards low-field. The comparison of shift values for vinylpyrazoles XVIII and XXI, XX and XXIV shows that in 5-methylsubstituted 1-vinylpyrazoles, the ring protons as well as H_A and H_X protons of the vinyl group undergo shielding, while H_B proton signals are low-field shifted. Deshielding value $\Delta\delta_B$ for both pairs are almost identical in CCl_4 and $CDCl_3$ (0,15 – 0,19 ppm), but decreases to 0,02 ppm in CD_3OD. $\Delta\delta_B$ Values for same pair of 1-vinylpyrazoles, found from the spectra of pure liquids, are equal to 0,15 and 0,19 ppm, respectively. The decrease of $\Delta\delta_B$ in CD_3OD is caused by a higher ability of the alcohol to association with N-2 pyrazole atom. Therefore, the effect of lone pair on H_B proton becomes almost negligible. The aforementioned low-field shift of H_B proton signals allows to draw a conclusion on predominant S-*cis*(N-2)-orientations of the vinyl group in 5-methyl-1-vinylpyrazoles. It is reported [187] that 2-alkylsubstituted 1-vinylpyrroles have the same structure.

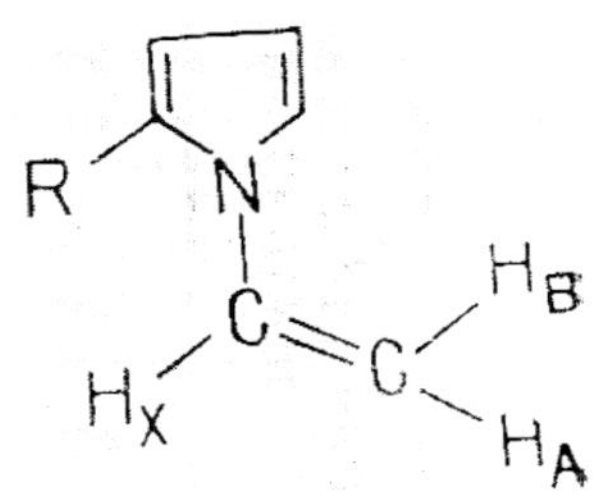

The additional information on relative orientation of the vinyl group and azole cycle has been obtained by comparison of spectral parameters of pyrrole and imidazole vinyl derivatives. It is found that 5-substituted vinylpyrazoles do exist mainly in S-*cis*-form, whereas the same compounds bearing no substituents in the position 5, represent a mixture of spatial isomers. The analysis of the PMR spectra shows that δ_B and δ_X values are rather informative in structural researches of vinylazoles containing heteroatom with lone electron pair, spatially accessible for N···CH coupling.

According to the current ideas of the NMR spectroscopy (see [187, 188] and the references cited therein), among the criteria of conjugation in vinyl compounds are heminal constants $\boldsymbol{J}_{AB}$ and chemical shifts of C_β atom signal of the vinyl group. From these positions the results of ^{13}C NMR spectral analysis using the method of paramagnetic additives are absolutely unexpected (Table 2.4). The dependence of $\boldsymbol{J}_{AB}$ constant values on substituent in the pyrazole ring has been compared.

The comparison of δ values of the ring carbon signals in the ^{13}C NMR spectra of unsubstituted pyrazole XXVI and its derivative XVIII shows that under the influence of the vinyl group C-3 and C-4 atoms are deshielded by 6,49 and 1,84 ppm, while C-5 atom is shielded by 6,95 ppm. The introduction of CH_3 group in the position 3 leads to shielding of the ring carbons, but the greatest high-field shift (2,7 ppm) is observed for the signal of the vinyl group C_β atom that is in agreement with electron effect of the conjugated system. Bromine atom at C-4 deshields C_β atom. In the spectra of 5-methylsubstituted vinylpyrazoles, signals of C_β atoms are low-field shifted (0,85 and 0,88 ppm) that points to slight decrease of conjugation in the molecule.

It is shown that the presence of substituent in the position 2 of 1-vinylpyrrole moiety leads to deviation of the vinyl group from the cycle plane and, as a result, to decrease of conjugation degree of nitrogen atom with the vinyl group. The reduction of the conjugation degree in 5-methyl-1-vinylpyrazoles can also be caused by violation of the molecule coplanarity. The increase of J_{AB} constant by 0,7 Hz in vinylpyrazoles XXII-XXV in comparison with compounds XVIII-XXI as well as low-field shift of C_β atom signals should testify to decrease of conjugation degree in molecules after introduction of the CH_3 in the position 5 of the cycle.

However for 1-vinylpyrroles, the increase of coplanarity is accompanied by both growth of J_{AB} value and the expected low-filed shift of the H_A proton signal. On the contrary, introduction of the methyl group in the position 5 of vinylpyrazoles, almost does not change the H_A proton signal. In vinylimidazole XXII, the increase of J_{AB} constant value by 0,8 Hz as compared to indazole XXI on can be considered increase in a constant as an indicator of reduction of extent of interface, but it contradicts the conclusions received in the analysis of values of signals ^{13}C.

Table 2.4. ^{13}C NMR spectral parameters of 1-vinylpyrazoles

Compound	δ, ppm					
	C-3	C-4	C-5	C_α	C_β	CH_3
XVIII	140,79	107,04	127,35	133,11	99,53	-
XIX	141,09	95,11	127,23	132,61	100,28	-
XX	148,91	105,74	127,22	131,77	96,83	12,23
XXIII	140,04	106,53	139,18	129,68	100,38	10,63
XXIV	147,78	104,94	137,08	127,69	97,71	8,85 (5) 11,62(3)
XXV	145,28	130,50	137,49	126,50	104,45	12,03(3) 8,97(5)
XXVI	134,30	105,20	134,30	-	-	-

Note: Positions of methyl moiety in the ring are given in brackets.

Inconsistency of the parameters, which reflect the effects of conjugation, allows to assume that value of constant can be caused by other factors. So, in [189] it was emphasized that heminal constants are tolerant to the influence of the substituent effect transmitted through orbitals and reflect induction effect. But in that case, the effects of 1-vinylimidazole, 3,5-dimethylpyrazole, 3,5-dimethyl-4-nitropyrazole and 5-methylpyrazole radicals will be identical, since in all vinyl derivatives J_{AB} values are practically the same.

It is known that in nitrogen-containing compounds, the values of heminal constants $\boldsymbol{J}_{HH}$ strongly depend upon the location of lone electron pair of nitrogen atom. Notably, high overlapping of electronic clouds of the lone pair with the C-H bond leads to higher value of $\boldsymbol{J}$ constant [190]. Since S-*cis*-(N-2) conformation is preferable for 1-vinylimidazole and 5-methylsubstituted vinylpyrazoles (due to steric hindrances), it is possible that in these compounds the increase of $\boldsymbol{J}_{AB}$ value is explained not only by alteration of conjugation in the molecule, but also by the effect of nitrogen lone electron pair. In that case, the irregularity of $\boldsymbol{J}_{AB}$ values alteration in 1-vinylazoles, depending on number and position of nitrogen atoms in the cycle, becomes clear [186]. A possibility of $\boldsymbol{J}_{AB}$ values application for assessment of electronic effects in vinyl derivatives of 1,2-azoles is reduced to establishment of the conformational homogeneity the compared compounds as well as to elucidation of the reasons defining the observed values of these constants.

It is a common knowledge that long-range spin-spin coupling constants $\boldsymbol{J}$ very often can serve as the independent tool in structural studies of the compounds, in which NMR spectra these constants are observed [191]. In this line, it is intriguing to check the applicability of this method to azoles [192]. It is natural that any new values of long-range spin-spin coupling constants are important for understanding mechanisms of such coupling transfer. The above circumstances stimulated the authors of the work [192] to analyze long-range spin-spin coupling constants $\boldsymbol{J}$ in 1-vinylpyrazole. Apparently, in the mentioned work, it has been reported for the first time on long-range spin-spin coupling in vinylazoles. PMR spectra of the vinylpyrazole vacuumized sample have been recorded under the conditions of decoupling from ^{14}N nucleus.

The investigations show that all protons of the heterocycle interact with the vinyl group protons. However, due to similar values of chemical shifts of H-3 and H-5 proton signals, the attempts to unambiguously determine values of all long-range spin-spin coupling constants fail. Substitution of hydrogen atom in the position 4 by bromine should simplify a shape of the spectrum without changing (at least, significantly) the spatial structure. Therefore, the spectrum of 1-vinyl- 4-bromopyrazole, recorded under the same conditions as the spectrum of vinylpyrazole, has been analyzed. The spectrum of 4-bromosubstituted derivative of VPZ is characterized by the following parameters ($\boldsymbol{\delta}$, ppm; $\boldsymbol{J}$, Hz): $\boldsymbol{\delta}_5 = 7{,}63$; $\boldsymbol{\delta}_3 = 7{,}54$; $\boldsymbol{\delta}_{\grave{A}} = 4{,}86$; $\boldsymbol{\delta}_{\hat{A}} = 5{,}45$; $\boldsymbol{\delta}_{\tilde{O}} = 6{,}97$; ${}^2\boldsymbol{J}_{AB} = 1{,}50$; ${}^3\boldsymbol{J}_{AX} = 8{,}90$; ${}^3\boldsymbol{J}_{BX} = 15{,}70$; ${}^4\boldsymbol{J}_{35} = 0{,}62$; ${}^5\boldsymbol{J}_{3X} = 0{,}32$; ${}^4\boldsymbol{J}_{5X} = 0{,}32$; ${}^6\boldsymbol{J}_{3A} = 0{,}43$; ${}^6\boldsymbol{J}_{3B} = 0{,}29$. Indexes at $\boldsymbol{\delta}$ and $\boldsymbol{J}$ correspond to designations on the scheme.

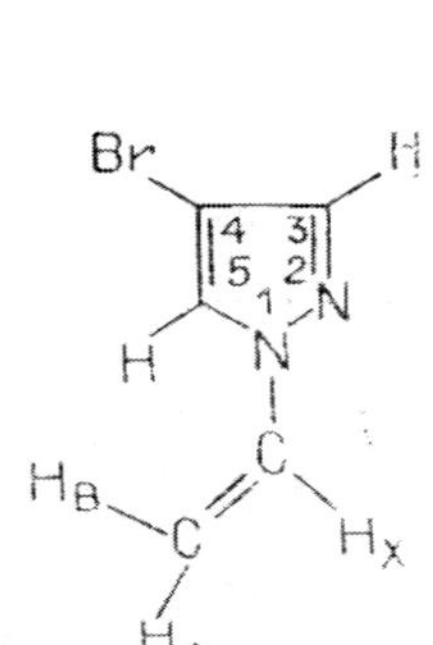

Assignment of the proton signals at the double bond is made according to δ and J values, characteristic for the PMR spectra of vinyl compounds. The key point here is identification of the ring proton signals, which has been performed using paramagnetic additive of $Gd(FOD)_3$.Under the corresponding conditions, the broadening caused by the paramagnetic complex is connected with distance r from the coordination ion to resonating nucleus and defined by the ratio: $\Delta \sim r^{-6}$ (see section 2.1 of this monograph).Therefore, transformation of the spectrum with the paramagnetic additive should lead to significantly different broadening of the signals, $\Delta_3 > \Delta_5$. Indeed, the addition of $Gd(FOD)_3$ to the sample results in the fact that one of the signals in a low-field, caused by the heterocycle protons, becomes considerably broadened in comparison with other signals.

Figure 2.27 shows high resolution PMR spectra (without the paramagnetic additive) after its interpretation, characterizing long-range spin-spin coupling in 1-vinyl-4-bromopyrazole.

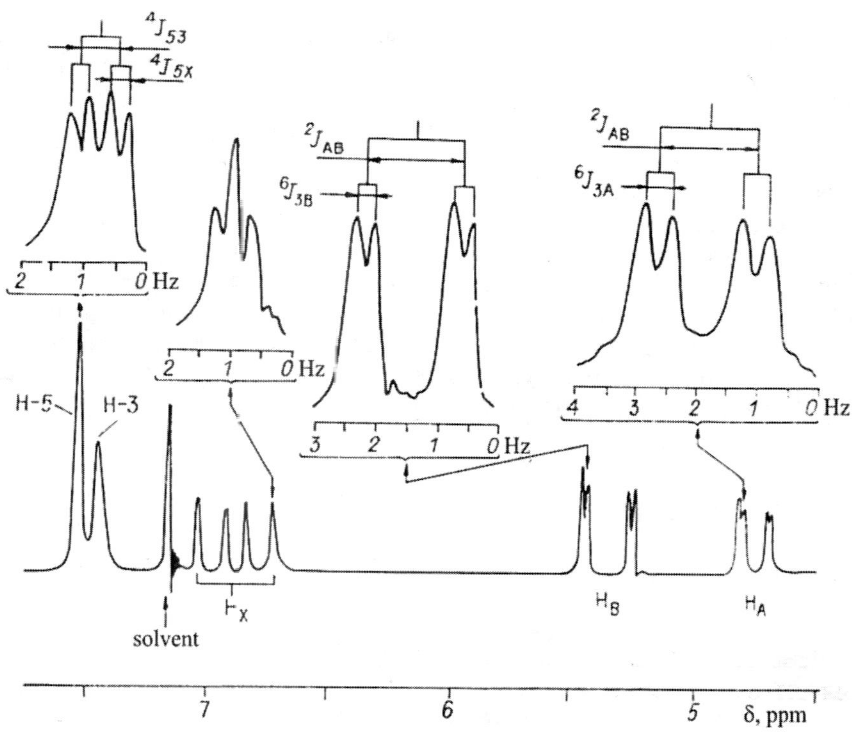

Figure 2.27. The PMR spectrum illustrating character of long-range spin-spin coupling in 1-vinyl-4-bromopyrazole.

From the PMR spectrum follows that long-range spin-spin coupling is realized between the cycle proton H-3 and the vinyl group proton H-4 via six bonds. This coupling is similar to that of the observed in N-vinylpyrroles [188].

According to the current literature representations, the discussed long-range spin-spin coupling is transferred mainly through plane zigzag-like fragment (Scheme A), i.e. it is realized when H-3 and H_A protons are of *trans*-configuration. This testifies to effective population of S-*trans*-(N-2)-conformations. At the same time, there is a long-range spin-spin coupling constant through five bonds between the cycle proton H-3 and the vinyl group proton H_X (0,32 Hz), which is also transferred via a plain zigzag-like fragment (Scheme B) [191].

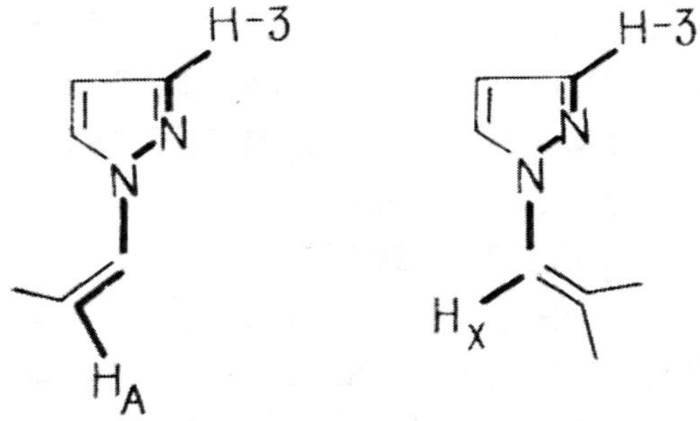

A Б

Therefore, S-*cis*-(N-2)-conformation of 1-vinyl-4-bromopyrazole is also populated. Thus, the analysis of long-range spin-spin coupling evidences the rotational *cis-trans*-isomerism in 1-vinylpyrazoles that confirms the conclusions [183] drawn independently on the basis of the analysis of chemical shifts in the PMR spectra.

In conclusion, the application of the paramagnetic additives method has allowed one to unambiguously identify ^1H and ^{13}C NMR spectra of a series of 1-vinylpyrazoles; to establish the peculiarities of spectral information of these compounds; to elucidate the nature of electronic effects of substituents introduced into the heterocycle; to solve the problem of *cis-trans*-isomerism of the mentioned compounds.

2.3.4. Vinyloxyphenylazomethines

The analysis of the NMR spectra of vinyloxyphenylazomethines is another example of high importance of the spectral information obtained with a help of the paramagnetic additives method. Though vinyloxyazome this have been synthesized in the sixties of the twentieth century, before work [193] there were no detail structural studies of these compounds. Such studies have been performed on the basis of ^1H and ^{13}C NMR spectra. The analysis of spectral information allows to conclude that these compounds exist in the form of *E*-isomers relative to both azomethine bond and olefinic C=C bond between the azomethine group and aromatic or heterocyclic ring. For example, N-(2-tenyliden-2-furylmethine) is characterized by S-*cis*(X, N)-conformation (X = S, O). The essence of the study is discussed below. Vinyloxyphenylazomethines can exist in the form of two isomeric structures:

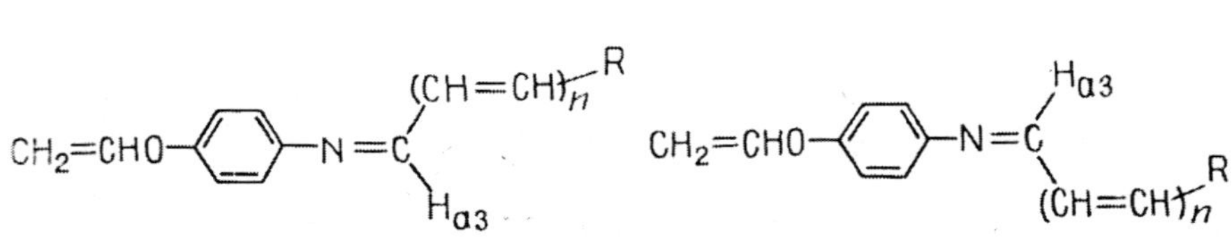

E- form Z-form

R = C_6H_5, 2-thienyl, 2-furyl ; n = 0,1.

The specified isomeric forms differ in a relative spatial disposition of the lone electron pair of the imine nitrogen and azomethine proton. Direct constants of $^{13}C - ^{1}H$ spin-spin coupling are known to be stereospecific relative to lone pair of the heteroatom that can be used for the identification of isomeric forms of azomethine compounds. The measured $^{13}C - ^{1}H$ constants with participation of the azomethine proton testify to the absence of additional positive contribution to their value due to the through-space effect of lone pair of imine nitrogen that should lead to *E*-isomer. Apart from isomerism relative to the C=N bond, azomethines with $n = 1$ can have also isomerism relative to the CH= CH bond.

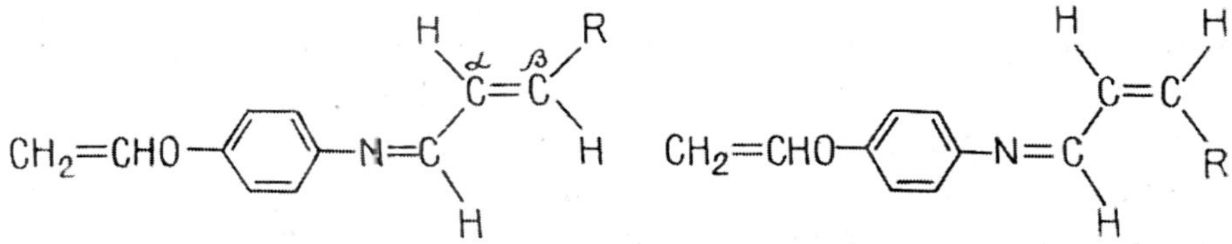

E- form Z- form

The type of isomerism in this case can be determined by the value of vicinal constants of spin-spin coupling between olefinic protons. ^{1}H and ^{13}C NMR spectra of vinyloxyazomethines with n=1 containing CH=CH bond between the cycle and azomethine group are quite complicated. In the PMR spectra of these compounds, signals of the vinyloxy moiety proton and azomethine proton are unambiguously identified. Signals of olefinic protons of the CH=CH bond "are hidden" by SN in the region of aromatic and heteroaromatic proton resonance (7,00 – 7,50 ppm). The presence of this olefinic fragment is proved by the data of ^{13}C NMR spectra.

LSR $\boldsymbol{Dy(FOD)_3}$ has been employed for the simplification of the PMR spectra of compound with $\boldsymbol{R} = C_6H_5$ – and assignment of the signals in these spectra. The addition of $\boldsymbol{Dy(FOD)_3}$ to a solution leads to considerable high-field shift of a signal of the olefinic proton located in $\alpha-$position relative to the azomethine moiety (Figure 2.28).Despite the essential paramagnetic broadening of this signal, its fine structure is detected. The signal of $\alpha-$olefinic proton is observed as a doublet of doublets due to the spin-spin coupling of this proton with azomethine ($\boldsymbol{J}_{\alpha a з}$ = 8 Hz) and $\beta-$olefinic proton ($\boldsymbol{J}_{\alpha\beta} = 16$ Hz). The latter constant corresponds to trans-disposition of the coupled protons that is realized in the formation of *E*-isomer.

It should be noted that application of $Eu(FOD)_3$ LSR, which shifts the signals towards low-field (relative to position in the spectrum of non-coordinated substrate), turns out to be inefficient, since in the mentioned field, other protons resonate, while the high-field fragment of the spectra of the studied azomethines is free. This example shows that in some cases, one should apply LSR, which shift the signals toward high-field (see section 2.4 of this monograph).

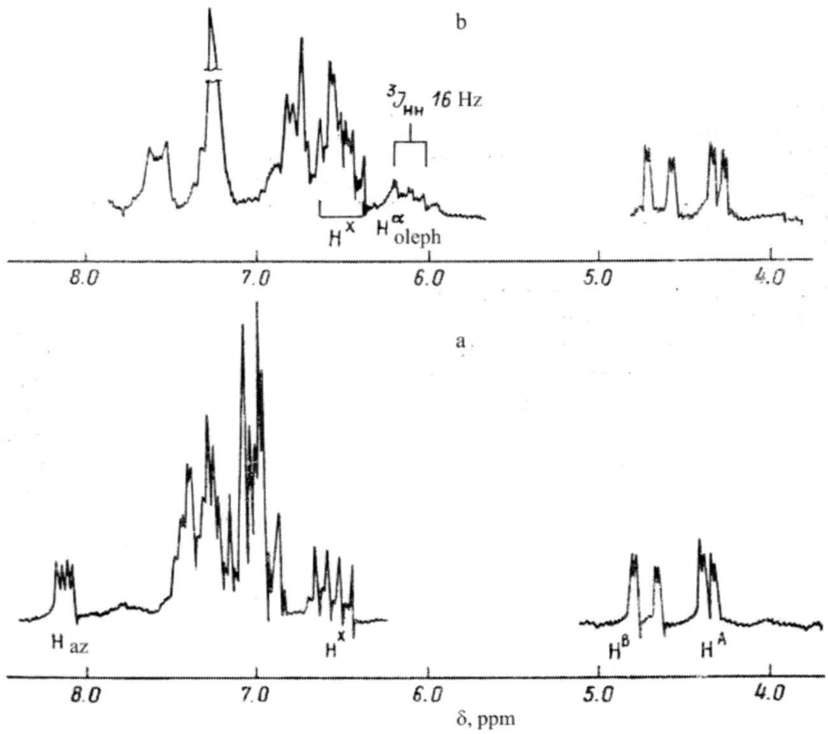

Figure 2.28. PMR spectrum of N-cynnamylidene-4-vinyloxyaniline without (a) and with (b) $Dy(FOD)_3$ paramagnetic additive.

2.3.5. 3- and 5-Substituted Flavones

Before publication of the work [194], literature contained scanty data on the PMR spectra of flavone and its 3- and 5-oxy- and methoxy- derivatives. In this connection, the PMR spectra of these compounds have been studied [194].

The ring protons A, B and C in the flavone molecule resonate in quite narrow region (6,5 – 8,2 ppm), thus complicating interpretation of their spectra. Assignment of signals of the methoxy and hydroxy group protons as well as H-3 proton (Figure 2.29) does not give rise to doubts.

The double resonance experiments show that upon suppression of the O – CH_3 group signal (Figure 2.29, curve 6), two doublets in the spectrum of 5-methoxyflavone at 6,71 and 6,98 ppm are assigned to the H-6 and H-8 protons, respectively. Such assignment is made taking into account the literary data on constants of long-range spin-spin coupling between the methoxy and aromatic protons. These constants should be higher for H-6 protons. Doublets of H-6 and H-6 protons in the spectrum of 5-oxyflavone have been assigned analogously.

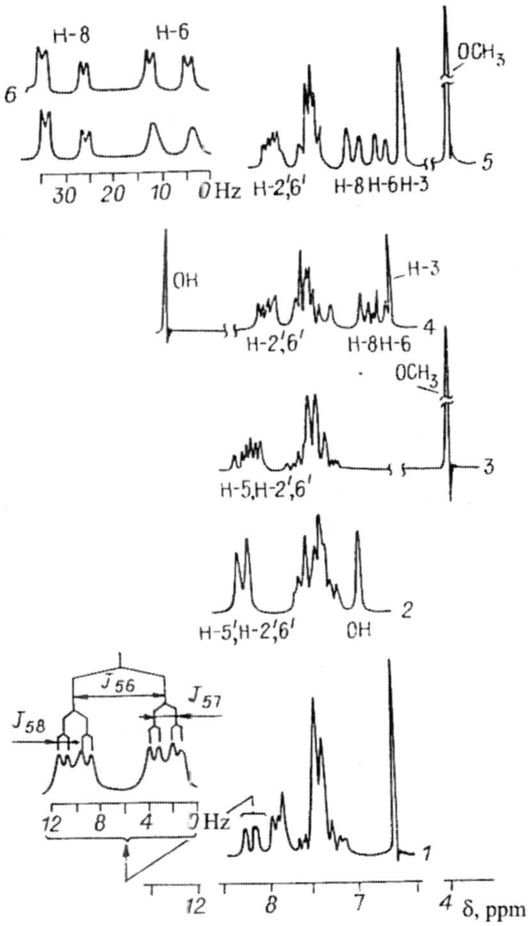

Figure 2.29. PMR spectra of flavones (1), 3-oxyflavone (2), 3-methoxyflavone (3), 5-oxyflavone (4), 5-methoxyflavone (5).

For detail interpretation of the PMR spectra, they have been recorded with additive of $Eu(DPM)_3$. Taking into account the data on basicity of flavone, it is assumed that the latter should be coordinated with $Eu(DPM)_3$ by the acyclic oxygen atom. Analysis of 3-methoxyflavone structure performed on Stewart models shows that signals of H-5 proton and methoxy group should be most significantly shifted. Therefore, the doublet in the low-field is assigned to H-5 proton (Figure 2.30, curve 2).

Thus, in the spectrum of 3-methoxyflavone recorded without the paramagnetic additive, this proton resonates in the low-field to give a doublet of quartets due to (like in the spectrum of flavones) spin-spin coupling with H-6, H-7 and H-8 protons. A signal of the H-5 proton, which is most close located to the coordination site, undergoes the highest broadening upon addition of $\boldsymbol{Eu(DPM)_3}$ to the studied sample. Therefore, splitting of this signal owing to the spin-spin coupling with H-7 and H-8 protons is not observed in the spectrum (see Figure 2.3. curve 2). Analysis of the PMR spectrum of 3-methoxyflavone, recorded with the paramagnetic additive and using additional radio frequency field corresponding to H-5 proton, has shown that the multiplet at 8,63 ppm is assigned to H-2 and H-6 protons. Signals of other protons in the PMR spectrum of 3-methoxyflavone recorded with additive of $\boldsymbol{Eu(DPM)_3}$ can be assigned by the double resonance technique. For example, a triplet is attributed to H-6 proton, while H-3, H-4, H-5, H-7, H-8 protons resonate in a low-field. Thus, the application of paramagnetic additives method allows increasing the informational content of the spectra of flavones and its 2- and 5-methoxy derivatives. Parameters of these spectra are summarized in Table 2.5.

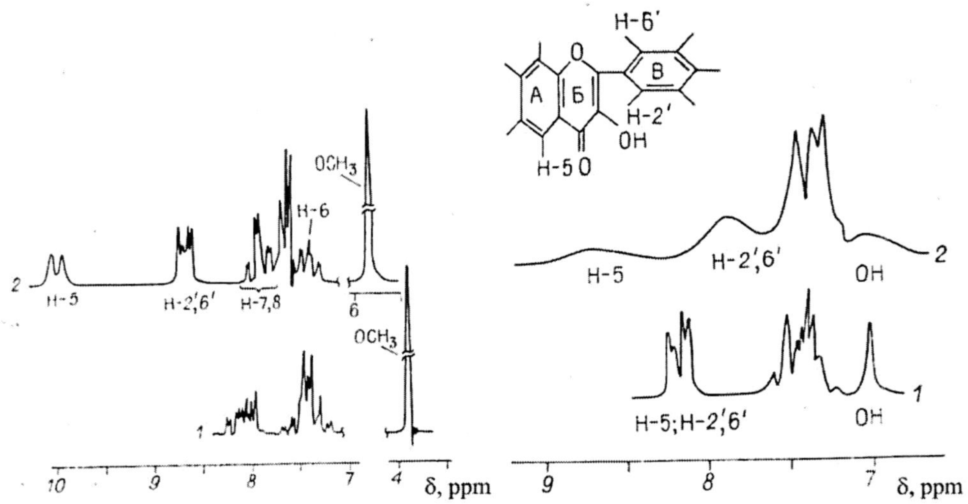

Figure 2.30. PMR spectra of 3-methoxyflavone recorded without (1) and with (2) additive of $Eu(DPM)_3$.

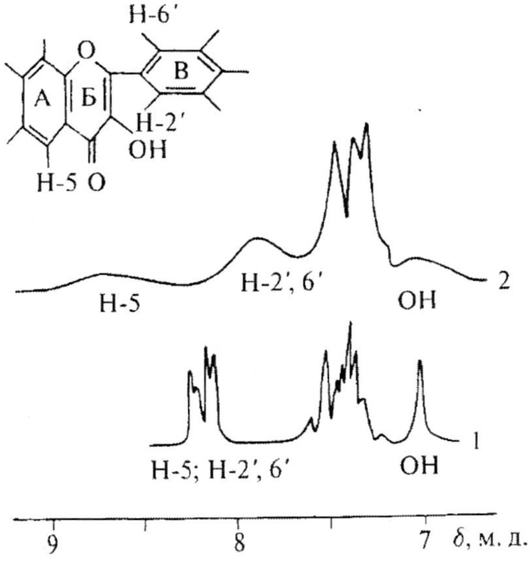

Figure 2.31. PMR spectra of 3-oxyflavone recorded without (1) and with (2) additive of $Eu(DPM)_3$.

Table 2.5. PMR spectral parameters of flavones

Compound	δ, ppm								J, Hz
	H-3	H-5	H-6	H-8	H-2′	H-6′	OCH_3	OH	
Flavone	6,81	8,23	-	-	8,00	8,00	-	-	$J_{56}=7{,}5$ $J_{78}=7{,}5$ $J_{57}=1{,}8$ $J_{68}=2{,}0$
3-Oxyflavone	-	8,19	-	-	8,19	8,19	-	7,01	-
3-Methoxyflavone	-	8,24	-	-	8,05	8,05	3,94	-	$J_{56}=7{,}5$ $J_{57}=1{,}7$ $J_{58}=0{,}7$
5-Oxyflavone	6,76	-	6,73	6,88	7,89	7,89	-	12,39	$J_{67}=8{,}5$ $J_{78}=8{,}1$ $J_{68}=1{,}0$
5-Methoxyflavone	6,49	-	6,71	6,98	7,80	7,80	3,90	-	$J_{67}=8{,}0$ $J_{78}=8{,}0$ $J_{68}=1{,}0$

The comparison of the paramagnetic shifts δ', induced by $Eu(DPM)_3$ additives in the PMR spectra of flavones, is of special interest. The δ' values are evaluated from the spectra recorded in the presence of the same amounts of ligand excess. In the spectra of flavone and its derivatives, the δ' values of H-2 and H-6 proton signals coincide. This is a consequence of averaging the induced shifts due to of free rotation around the C_2–C_1 bond. It is found that the $\delta'_5/\delta'_{2,6}$ ratio in the spectrum of flavone is more than six, while in the spectrum of 3-methoxyflavone, it is equal to three.

The $\delta'_3/\delta'_{2',6'}$ ratio for flavone is 4,5. In the spectrum of 5-methoxyflavone, it exceeds 5. In addition, in the spectra of 3- and 5-methoxyflavones, the same $\delta_{2',6'}$ values correspond to noticeably different $\delta'_{O\ CH_3}$ values. The signals of H-3, H-4 and H-5 protons in flavone and its methoxy derivative change almost equally upon addition of $Eu(DPM)_3$ into the studied samples. At the same time, δ' values of the ring A proton signals are different, and this distinction is expressed for flavone to a lesser extent.

Special attention is given to the fact that in the spectrum of 3-methoxyflavone, δ'_8 value is higher than δ'_7 and δ'_6 values. Comparison of the observed and expected (from the Steward's model values) paramagnetic shifts allows one to conclude that the above difference in δ' values is connected with various orientation of A, B and C rings in the studied flavone molecules.

Unfortunately, an attempt to analyze the PMR spectra of 3- and 5-oxyflavones using the above technique has met with no success, since the addition of $Eu(DPM)_3$ to the studied samples of these compounds does not lead to the expected shift of signals. The increase of $Eu(DPM)_3$ concentration in the sample results in strong broadening of the signals, the value of the broadening being comparable with the observed paramagnetic shifts. This is illustrated by the spectra shown in Figure 2.31. When $\mathbf{P}_\mathbf{M} << 1$ (this ratio is used in the experiments), the observed changes can be caused by two reasons: (i) the absence of complete averaging (slow exchange); in the paramagnetic complexes formed by 3- and 5-oxyflavones, relaxation time $\mathbf{T}_{\mathbf{2M}}$ of resonating nuclei is such that the small additive of $Eu(DPM)_3$ to the studied sample leads to considerable broadening of signals.

The right conclusion can be drawn from the dependences $\delta' = \delta'(t)$ and $\Delta = \Delta\,(t)$. The temperature measurements have shown that in the temperature range from – 20 to 80° C, for the observed paramagnetic shifts (except for δ of H-3 proton signal in the spectrum of 5-oxyflavone), it is possible to achieve complete averaging, but in this case, the lines remain also strongly broaden. According to the theory, expression for $1/T_{2M}$ can be presented as follows (see section 2.1 of this monograph):

$$1/T_{2M} = B'r^{-6}\tau_c + C'(\delta'_k)^2\tau_e \quad . \tag{2.28}$$

When the contact contribution is absent, the $1/\mathbf{T}_{\mathbf{2M}}$ value is defined by both the distance r from coordination ion to resonating nucleus and relaxation effects. It has already been mentioned that the paramagnetic ion Eu^{3+} is characterized by very short of electron relaxation time ($\tau_s \sim 10^{-13}$ c), therefore, $1/\tau_c$ and $1/\tau_e \approx 1/\tau_s$ .The obtained experimental data testify that in a complex of $Eu(DPM)_3$ with 3- or 5-oxyflavone, T_{2M} value appears to be significantly lesser than one could expect. From the expression (2.28) follows that the decrease in T_{2M} is due to the increase of τ_s value upon coordination or (and) origination of the contact contribution in paramagnetic shift. As for the first reason, it is associated with peculiarities of electronic structure of the paramagnetic ion. Indeed, according to the current theoretical representations, paramagnetism of Eu^{3+} is a consequence of thermal population of its first excited level ($\Delta\mathbf{E} = 360\ \text{cm}^{-1}$).

It has been demonstrated that for Eu^{3+}, even higher levels can be thermally populated [195]. Having this in mind, it is possible to assume that the coordination leads to the increase in efficiency of population of others (apart from $^7F_1$) excited levels. This should result in alteration of τ_s values and, as a consequence, alteration of relaxation times of nuclear spins. Notably, in this case, the delocalization of non-compensated electron spin density on the ligand, causing the appearance of contact shifts, does not take place. For differentiation of the above-described mechanisms leading to decrease of T_{2M} values of 3- and 5-oxyflavone nuclei and conformation of electron delocalization on the ligands, the relative paramagnetic shifts of H-2′, H-6′, H-3, H-5 proton signals have been compared. The difference in signs and values of the ratios can be considered as the evidence of the contact contribution to the shifts induced in $Eu(DPM)_3$ in the PMR spectra of 3- and 5-oxyflavones.

Compound	$\delta'_5/\delta'_{2',6'}$	$\delta'_3/\delta'_{2',6'}$
3-Oxyflavone	– 1,4	-
3-Methoxyflavone	3,1	-
5-Oxyflavone	-	1,5
5-Methoxyflavone	-	5,7

A possibility of the delocalization through the excited state of the substrate molecule coordinated with paramagnetic ion has been demonstrated [196]. Actually, we mean here the dative interaction S ← L leading to population of the excited levels of S molecule. In this case, the atomic orbitals, presented in molecular orbitals of the excited states, will contain unpaired spin density ρ, which induces the shifts ω_K values of ligand nuclear signals. The ω_K value depends on the constant of contact HFC, which can be expressed as follows [178]:

$$A = Q\frac{\rho}{2J}, \tag{2.29}$$

where $Q-$ is a constant for the corresponding $s-$orbital; $J-$ is a total spin of the complex. Absolute Q values reach hundreds and thousands oersted [178], therefore, even a several percent increase in population of the excited state of the ligands should lead to induction of considerable shifts.

Owing to structural peculiarities of 3- and 5-oxyflavones (the proximity of relative disposition of the carbonyl and oxy groups), one could assume that the observed transformation of the PMR spectra [196] results from the substitution of one or several DPM molecules bonded with Eu^{3+} by the molecules of these flavones, i.e. that the latter form with europium the same bonds as DPM. However, the experimental data do not confirm such assumption. Indeed, 3- and 5-oxyflavones form quite strong intramolecular bonds. This is confirmed by UV, IR and PMR spectra, for example, by the PMR spectrum of 5-oxyflavone (Figure 2.32), where the hydroxyl proton induces a narrow signal in the low-field (12,36 ppm). The position and width of this signal do not depend upon temperature of the studied sample.

Besides, if the substitution of DPM molecules would take place, the broadening of L and S proton signals upon changing the sample temperature and the concentration ratio [L]/[S] would be comparable. Actually, this is not the case. In the spectra of samples recorded with $Eu(DPM)_3$ additives, the signals of flavone protons are much more broaden in comparison with the signals from $t-Bu$ $Eu(DPM)_3$. Unusual transformation of the PMR spectra of 3-

and 5-oxaflavones is in agreement with the fact that these compounds are characterized by abnormal UV, IR and luminescent spectra [197]. In particular, energy of the first singlet transition in 3- and 5-oxyflavones is by approximately 0,3 eV lower. Besides, fluorescence owing to transition of $S_{\pi\pi}$ type is inherent in these flavones.

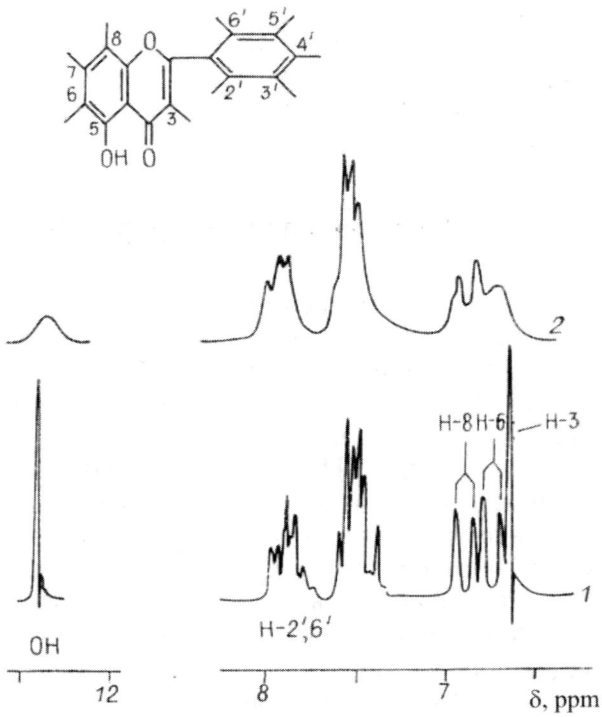

Figure 2.32. PMR spectra of 5-oxyflavone without (1) and with (2) additive of $Eu(DPM)_3$.

The reality of contact HFC transfer in the complexes of 3- and 5-oxyflavones with $Eu(DPM)_3$ through the excited states of ligand molecules allows one to predict a new effect, so-called chemically induced dynamic nuclear polarization (CNP) by paramagnetic additives. It is known (see, for example, [198]) that one of the reasons of CNP is a coupling between unpaired electron spins and nuclear spins, which signals are observed. Usually, HFC, occurring in the course of coordination, results in the signals shift and broadening. The nonequilibrium population of energy levels, which induces CNP, does not take place. If, upon coordination, the unpaired spin density is delocalized on vacant MO of the ligand, the molecule of the latter will be in a state with non-compensated electron spin even after transfer from the sphere of paramagnetic ion in the solution. This state has the corresponding lifetime; therefore, one should expect the appearance of the CNP effect in the NMR spectra.

It should be emphasized that the described unusual transformation of PMR spectra of 3- and 5-oxyflavones under the action of $Eu(DPM)_3$ represents a first example of such effect of LSR. The authors of a later publication [199], discussing the interaction of N-butyl-N-(4-methoxyazol-2-ol)-2-methylpropionamide with fluorine-containing LSR, came to the conclusion that unpaired spin density is localized on the heterocycle. This leads to unusually broaden ^1H signals and considerable paramagnetic shifts of ^{13}C signals of this cycle.

2.3.6. 2-, 3-, 4-Pyridylcarbinol Vinyl Ethers

The analysis of PMR spectra of these compounds has shown that despite the qualitative similarity, they have some distinctive features (Table 2.6) [200].

1. The resonance line of CH_2 group protons in the spectrum of ether XXVII is shifted by 0,14 ppm towards low-field as compared to the spectrum of ether XXVIII. Such shift can be resulted from different contribution caused by the effect of nitrogen magnetic anisotropy. However, first, such shift is less than one can expect and, second, the shift values of this group proton signals in the spectra of compounds XXVIII and XXIX are almost the same. The difference is in the position of CH_2 group proton signals is apparently conditioned not only by the effect of nitrogen magnetic anisotropy.
2. Upon transition from ether XXVII to ether XXIX, the resonance lines of H_A and H_B protons slightly change their positions in the PMR spectra, a signal of the H_A proton being shifted more significantly than that of the H_B proton. It is supposed (see section 2.3.3 of the monograph) that the signal positions of the vinyl group $\beta-$protons in the spectra of vinyl ethers is defined, in particular, by conjugation of oxygen atom to the double bond that reflects the internal chemical shift $\delta'' = \delta_B - \delta_A$. Therefore, the observed reduction of δ'' value upon transition from compound XXVII to XXIX can be a consequence of redistribution of electron density in the molecules.

Table 2.6. PMR spectral parameters of 2-, 3-, 4-pyridylcarbinol vinyl ethers XXVII – XXIX

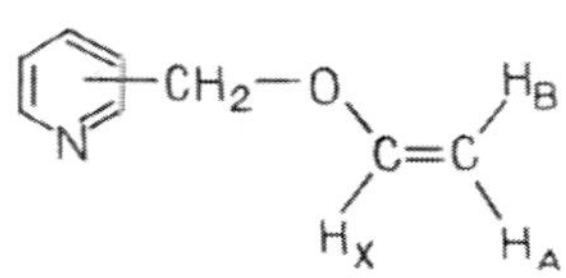

Compound	δ, ppm									J, Hz
	$\delta_{À}$	$\delta_{Â}$	$\delta_{õ}$	δ_2	δ_3	δ_4	δ_5	δ_6	δ_{CH_2}	
XXVII	3,96	4,21	6,45	-	7,30	7,55	7,03	8,40	4,80	$J_{56}=5,0$ $J_{43}=7,6$ $J_{54}=7,3$ $J_{46}=1,8$
XXVIII	3,98	4,16	6,43	8,41	-	7,51	7,11	8,38	4,66	$J_{42}=J_{46}=$ 1,3 $J_{45}=7,8$ $J_{56}=4,8$ $J_{52}=0,5$
XXIX	4,01	4,14	6,44	8,44	7,10	-	7,10	8,44	4,68	-

For more detail structural study of vinyl derivatives of pyridylcarbinol, $Eu(DPM)_3$ additives are used. It is assumed [178] that ether oxygen coordinates with LSR significantly weaker than the pyridine nitrogen. Therefore, it is postulated [200] that ethers XXVII-XXIX should coordinate via the nitrogen atom. Figure 2.33 shows the PMR spectra of the studied compounds recorded with and without $Eu(DPM)_3$ additive. Transformation of the vinyl ether spectra due to hyperfine coupling seems to be unexpected. Indeed, in the spectrum of compound XXVII (Figure 2.33, curve 2), the greatest shift is observed for signals of the CH_2 group and olefinic proton. The δ'_B and δ'_3 values are also approximately the same, but exceed the δ'_A value by two times. The signals of H-6, H-5 and H-4 protons are shifted slightly. Interestingly, δ''_6 values, unlike the δ' values of other signals, becomes negative (paramagnetic pseudo-contact shift towards low-field is positive). Paramagnetic shifts observed in the spectrum of compound XVII correspond to its coordination with $Eu(DPM)_3$ through the ether oxygen.

Unlike ether XXVII, in the spectrum of compound XXVIII, H-2 and H-6 proton signals have the highest shift values (Figure 2.33, curve 4), δ'_2 и $\delta'_6 > \delta'_5 > \delta'_4$, and δ'_{CH_2} is 1,5 times lower than δ'_5. The δ'_X and δ'_B values are practically the same and are by more than 10 times less than δ'_2 and δ'_6 values. H_A proton signal has the lowest δ' value. It should be noted that δ' values are different for H-2 and H-6 proton signals. Signal of these protons can be easily assigned, since one of them is a singlet (H-2), and other is a doublet owing to spin-spin coupling with the neighbor H-5 proton. Like in the previous case, in the spectrum of ether XXIX, H-6 and H-2 proton signals have the highest δ' values (Figure 2.33, curve 6). However, it is important to underline a peculiarity of this spectrum range: protons of the CH_2 and vinyl group show two signals each. The analysis of spatial structure of the ether molecules discussed here reveals that in compound XXVIII (due to the steric hindrances for specific conformation relative to the $C-CH_2$ bond), the pyridine ring can not be located in the plane ensuring the conjugation of the double bond $\pi-$system to oxygen p_z-electrons.

In ethers XXIX and XXVIII, other orientation of the heterocycle and vinyl group is possible. In the long run, this influences the character of conjugation in the molecule as well as the coordination with $Eu(DPM)_3$. As for 4-pyridylcarbinol vinyl ether, it demonstrates two energetically favorable conformations with different orientation of the pyridine ring and vinyl group. The signals of both conformations have identical value of chemical shifts, or the observed shifts are average. When vinyl ether is located into the coordination sphere of the paramagnetic complex, the substituent R in the ligand molecule has other orientation, since δ' is defined by distance to the coordination ion and orientation of resonating nucleus. This confirms irregular dependence of δ' values of different proton signals upon temperature of the studied sample. Moreover, the change in temperature leads to alteration of the integral intensities ratio of the conformer signals. Actually, the effect of $Eu(DPM)_3$ is to increase the chemical nonequivalence of CH_2 and the vinyl group resonating nuclei so that at given rate of conformational transitions, the signals of each conformer will be observed in the spectra.

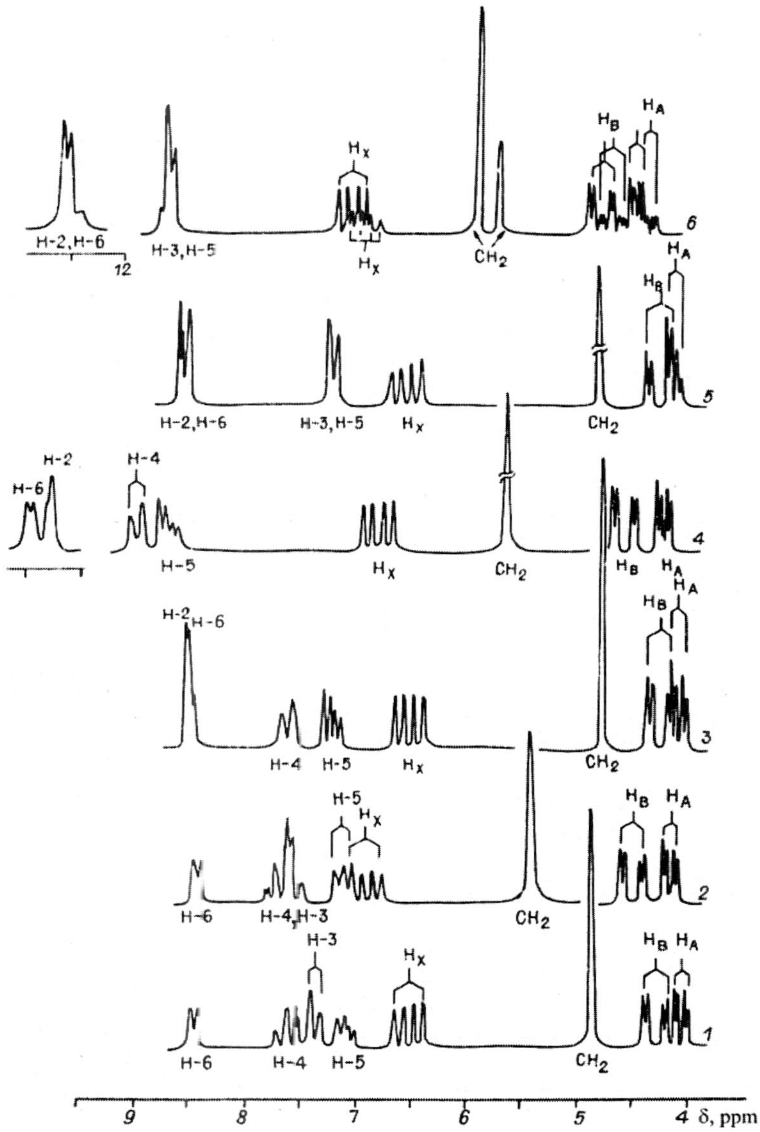

Figure 2.33. PMR spectra of 2, 3, 4-pyridylcarbinol vinyl ethers without (1, 3, 5) and with (2, 4, 6) additive of $Eu(DPM)_3$.

A property of LSR to considerably increase the chemical nonequivalence of resonating nuclei belonging to different conformations of the same molecule has been successfully employed by S.P. Babaylov for the study of nature of intramolecular chemical exchange in complexes of 18-membered macrocyclic polyethers with lanthanide cations, and also for the calculation of the activation parameters characterizing these processes [201].

2.3.7. 2,5-Dimethyl-4-Methylene-1,3-Oxathiolane.

Table 2.7. PMR spectral parameters of 2,5-dimethyl-4-methylene-1,3-oxathiolane isomers

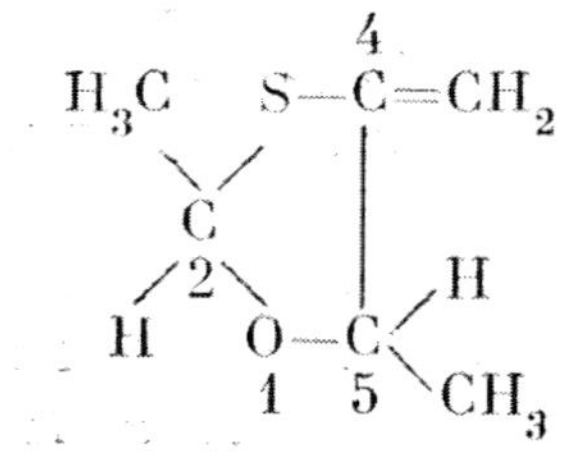

Isomer	δ, ppm					J, Hz
	CH_3-2	CH_3-5	H-2	H-5	$=CH_2$	
cis	1,48	1,33	5,27	4,30	4,82	$J_{2-CH_3} = 5{,}8$ $J_{5-CH_3} = 6{,}1$ $(J_{46} + J_{45})/2 = 2{,}1$
trans	1,46	1,25	5,46	4,75	4,82	$J_{2-CH_3} = 5{,}9$ $J_{5-CH_3} = 6{,}4$ $(J_{46} + J_{45})/2 = 1{,}6$

The interaction of sodium sulfide with acetylene in the presence of water affords new oxygen- and a sulfur-containing heterocycles, the structure of which has been proved by a series of techniques including the NMR spectroscopy [202]. In the PMR spectrum of this compound, signals of the methine protons appear as quartets due to the spin-spin coupling. Notably, the proton attached to carbon, bonded with oxygen and sulfur atoms, resonates in the weakest field. The methyl protons are responsible for the emergence of signals in high-field. In the PMR spectrum, the presence of four quartets from two methine protons and eight doublets from the methyl group protons indicates the existence of two isomers in the sample which are caused by relative disposition of the CH_3-groups. The signals of two olefinic protons are overlapped to give unresolved multiplet with the center at $\delta =$ 4,82 ppm. Character of proton signals at the double bonds testifies that the cycle is a five-membered one. In the case of six-membered cycle, the signal of olefinic proton (relative to the heteroatom) should be shifted significantly towards a weaker field as compared with the signal of $\beta -$ protons. From the spectrum, it follows that one of the methine protons couples with protons at the double bond. As a consequence, quartet lines are split into triplets, and the constant of this spin-spin coupling in isomers become different.

To assign the signals of each of these isomers, the spectra of samples have been recorded with additives of $Eu(DPM)_3$. These spectra are characterized by the following parameters: a signal of the methine and methyl protons of each isomer is shifted equally; δ' values of the isomer signals differ approximately by 3 times. Obviously, in the case of *cis*-orientation of the CH_3-group, one should expect almost the same δ' values of these groups' signals. Signals of the methyl groups should also have the similar δ' values. In the *trans*-isomer, owing to steric hindrances, the coordination conditions are different from the *cis*-isomer that is confirmed by the triple distinction in values of experimental paramagnetic shifts of the isomers. Thus, the signals with higher δ' values should be attributed to the *cis*-isomer. Taking this into account and according to results obtained by other methods of NMR spectroscopy, it has been established that the studied compound is 2,5-dimethyl-4-methylene-1,3-oxathiolane (Table 2.7).

2.3.8. Configuration and Conformation of 8-(3,4-dihydropyranyl-2) Oxyquinoline Derivatives

This section deals with the PMR spectral analysis of the structure of 8-(3,4-dihydropyranyl-2) oxyquinolines XXX – XXXII [203].

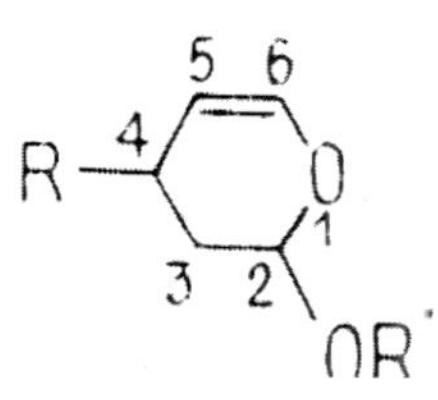

XXX R=H
XXXI R=CH_3
XXXII R=C_6H_5

R'=

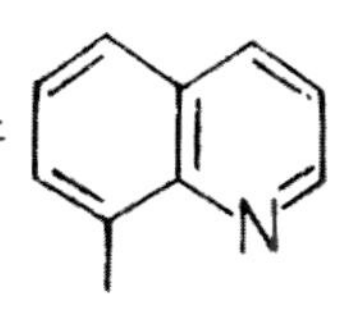

In the PMR spectra of compounds XXX-XXXII, signals of all protons of the dihydropyrane cycle (and the pyridine ring, when R = C_6H_5) are doubled that indicates that both products exist as a 1:1 mixture of stereoisomers. In usual spectra, owing to overlapping of the signals of different isomers, it is impossible to unambiguously determine the chemical shifts and spin-spin coupling constants. This difficulty has been overcame, when the spectra of the samples have been recorded with additive of $\boldsymbol{Eu(DPM)_3}$.

The addition of LSR results in the increase of relative values of the chemical shifts that allows the double homonuclear resonance technique to be applied for unambiguous determination of the PMR spectral parameters (δ and J). Some of these parameters, most representative (from the viewpoint of the problem investigated here), related to the pyrane cycle are given in Table 2.8. The table contains also $\boldsymbol{n}$ values, which will be discussed below.

To establish configuration and conformation of the isomers, one should account for the stereochemical dependences of vicinal, allylic and long-range spin-spin coupling constants. Orientation of the substitutent in the position 4 of the dihydropyrane ring is evaluated from the values of allylic constants ${}^4J_{46}$, which equal to 2,0 and 2,5 Hz for isomers of compound XXXII. Such constants can be achieved, if the value of φ angle is close to zero (according to literature data):

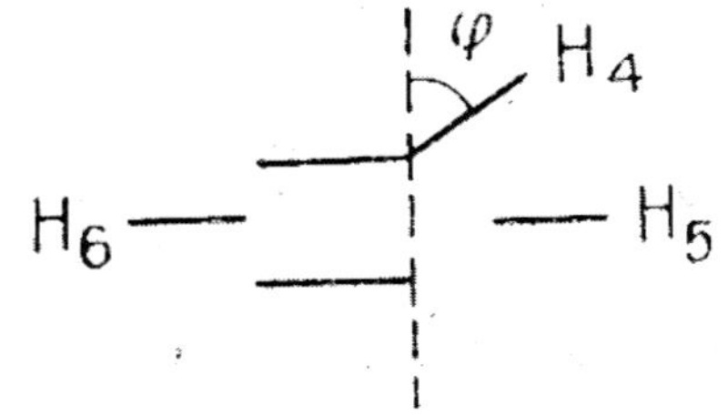

This indicates the equatorial (or pseudo-equatorial) orientation of the phenyl group. The methyl moiety at C-4 atom in the isomers of compound XXXII is also located in the equatorial position. The signal of H-2 proton in compound XXX is presented by a triplet, while in the spectra of derivative XXXII, these protons appear as multiplets, namely triplet and quartet. Since the equatorial configuration of the substituent in the position 4 has been established for both isomers of compound XXXII, the triplet is assigned to *trans*-isomer, and the quartet is attributed to *cis*-isomer. The spectrum of compound XXXI is similar to that of derivative XXXII.

The values of vicinal constants $^3J_{23}$(see Table 2.8) allow to assume that in dihydropyranes XXX and trans-isomers XXXI, XXXII, the distinct anomeric effect is observed because $^3J_{23}$= 2,5 Hz corresponds to equatorial-axial interactions, i.e. 8-quinolineoxy group is in the axial position. The values of vicinal constants $^3J_{23}$for *cis*-isomers XXXI and XXXII testify that in this case, either equilibrium mixture of conformers is formed, or the predominant conformation is the simplified semi-chair. The $^3J_{23}$value of the substituent at C-4 atom defines, in the first case, the position of conformational equilibrium, and, in the second case, degree of planarization of the semi-chair conformation. The second of these assumptions is not confirmed by temperature dependence of the spectrum. The increase of the sample temperature up to 75° C leads to reduction of the $^3J_{23}$ value by 1 Hz. This corresponds to 10% augmentation of the axial conformation fraction. When temperature decreases, equilibrium is shifted towards the equatorial conformer. The application of "average" standard values of J_{aa}, J_{ae}, J_{ee} make it possible to estimate the position of the conformational equilibrium, which is characterized by population n of the axial conformer. The results of this evaluation are as follows (see Table 2.8):

1) Compounds XXX, XXXIIa have the conformation with axial aroxy group;
2) *Cis*-isomers represent an equilibrium mixture of conformers, obviously, owing to counteraction of the anomeric effect and steric hindrances in the axial conformation.

It may appear that the values of allylic constants $^4J_{46}$contradict the second assumption, since these values in XXXIb and XXXIIb do not correspond to the fraction of the axial conformer for each of these compounds. This difficulty can be obviated, if to assume that the cis-isomers in the dihydropyrane cycle have not "semi-chair" conformation, but "sofa" conformation with plane disposition of atoms in $C_3 - C_4 - C_5 - C_6 - O$ fragment of the dihydropyrane cycle. Such conformation satisfies to values of all constants of the spin-spin coupling. Inversion of the conformations in this case occurs during rotation around the O – C_2 and C_2 – C_3 bonds:

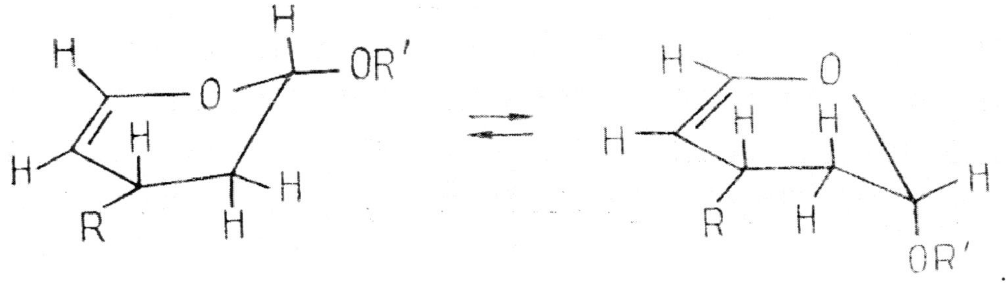

As for the *trans*-isomers, it is unjustified to believe that they may not have "semi-chair" conformation, which is inherent in cyclohexane and dihydropyrane systems.

A shape of the H-5 proton signal in the spectra of quiolines XXXI and XXXII can be considered as an independent confirmation of the above conclusions on configuration and conformation of the compounds studied. For *trans*-isomers of XXXIa and XXXIIa, this signal appears as a doublet ($J_{56} = 6{,}0$ Hz), each component of which is split in the triplet, which is caused by the coupling with H-4 (2,0 Hz) and H3e (2,0 Hz) protons (Figure 2.34). Steric dependence of the vicinal constant of spin-spin coupling shows that J_{45} is characteristic for dihedral angle = ~ 90°, i.e. it confirms axial position of the proton and equatorial position of the substitutent at C-2 atom. The coupling of H-5 proton with H-3e proton is transmitted through four σ – bonds, and in non-conjugated cyclic systems, it is usually realized through plain zigzag-like fragment of W shape.

This indicates the existence of predominant conformation of "semi-chair" type. In the *cis*-isomers of compounds XXXIb and XXXIIb, each component of H-5 proton is not a triplet, but a doublet of doublets owing to the decrease of $^4J_{35}$ constant (Figure 2.34). Such type of the signal evidences that the long-range spin-spin coupling of H-5 proton is possible only with one of H-3 protons. The study of a spatial model shows that in the case of realization of "sofa"-conformation in the dihydropyrane cycle, inversion of the conformations leads to the situation, when the plane W-shaped positions of H-3 and H-5 protons will be observed only for one of H-3 protons, while the second proton is not coupled with H-5. This conclusion is also supported by the decrease of long-range spin-spin coupling constant in *cis*-isomers. A higher symmetry of the "semi-chair" conformation results in the following. Due to the cycle inversion, H-3 protons are sequentially take the equatorial position and should cause the triplet splitting of H-5 signal in the spectrum of *trans*-isomer.

Thus, the results obtained have shown that the *trans*-isomers of the compounds studied have predominantly "semi-chair" conformation and exert the anomeric effect. For labile *cis*-isomers, the "sofa" conformation is realistic.

Table 2.8. PMR spectral parameters of 8-(3,4-dihydropyranyl-2) oxyquinoline

Compound	R	Isomer	δ, ppm					J, Hz					n %
			δ_2	δ_3	δ_4	δ_5	δ_6	J_{23} a	J_{23} e	J_{45}	J_{46}	J_{35}	
XXX	H	-	6,04	-	1,5 2,8	4,87	6,14	2,5	2,5	-	2,0 1,2	-	100
XXXIa	CH_3	*trans*	5,96	1,57 2,30	2,3	4,71	6,10	2,5	2,5	2,0	2,4	2,0	100
XXXIb	CH_3	*cis*	5,86	1,99 2,30	2,3	4,56	6,15	6,5	2,5	2,4	1,9	1,0	53
XXXIIa	C_6H_5	*trans*	5,99	1,95 2,53	2,5	4,90	6,20	2,5	2,5	2,0	2,5	2,0	100
XXXIIb	C_6H_5	*cis*	5,89	2,24 2,53	2,5	4,70	6,30	8,5	2,5	2,0	2,0	1,2	30

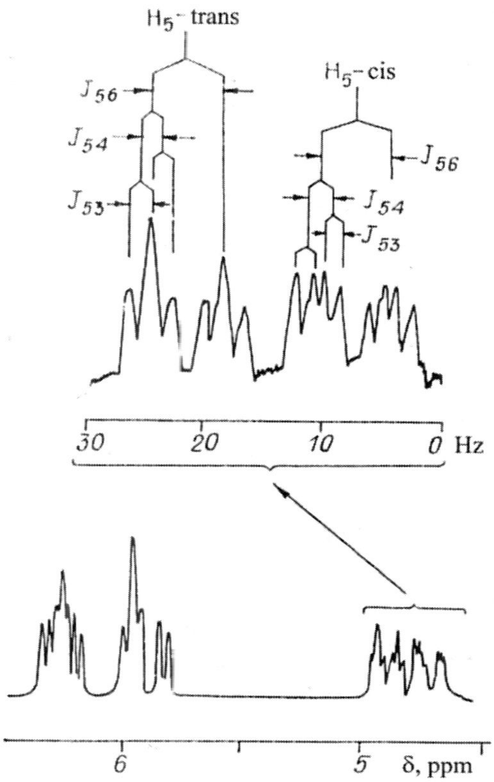

Figure 2.34. A fragment of the PMR spectrum of compound XXXII (a mixture of *trans*- and *cis*-isomers).

2.3.9. 2H,6H-2,6-Dimethyl-4-Amino-1,3,5-Dithiazine and Its Derivatives

In some cases, structural information can be obtained by combination of the complex-forming additives method with other qualitative approaches. In this line, we can mention here the work [204] dealing with the structural study of 2H,6H-2,6-dimethyl-4-amino-1,3,5-dithiazine XXXIII and its derivatives XXXIV-XXXVII. The PMR spectra of these compounds have been recorded in diverse solvents and transformed by the paramagnetic additives. The key point of this study is the problem of predominant tautomeric form, in which the aforementioned compounds can exist.

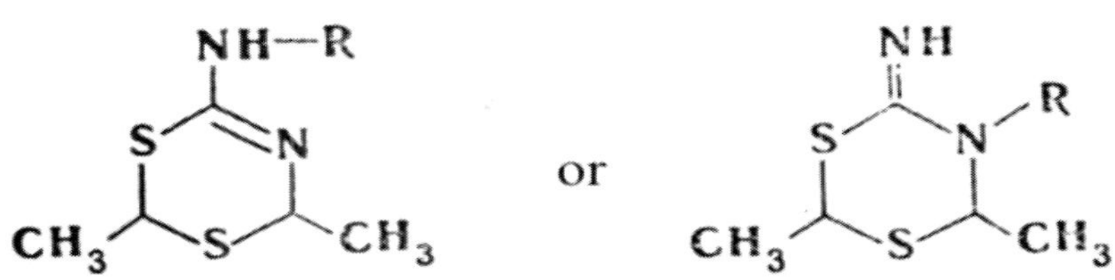

XXXIIIa,b R=H; XXXIVa,b R=$COCH_3$; XXXVa,b R=COC_3H_7;
XXXVIa,b R=COC_6H_5; XXXVIIa,b R=$SO_2C_6H_4CH_3$

First of all, it has been assumed that different (in properties) solvents exert different effects on tautomeric equilibrium and on the parameters of the NMR spectra of a mixture of tautomers (amine *a*and imine *b*).

Table 2.9 shows the chemical shifts of proton signals of the methine and methyl groups of the studied heterocycles, the values of J_{CH-CH_3} constants as well as δ values of proton signals coupled with the nitrogen atom. The data of Table 2.9 testify that the replacement of CCl_4 by CD_3OD and DMSO-D_6 influences the δ value of compound XXXIII, but only in limits, which usually are defined by the alteration of nature of intermolecular interaction between substance and solvent. In its turn, J_{CH-CH_3} value remains unchanged.

Table 2.9. PMR spectra of compounds XXXIII – XXXVIII

Compound	Solvent	δ ,ppm			J_{CH-CH_3}, Hz	Compound	Solvent	δ ,ppm			J_{CH-CH_3}, Hz
		CH	CH_3	NH				CH	CH_3	NH	
XXXIII	CCl_4	4,38 4,60	1,50 1,54	4,34	6,6	XXXV	CCl_4	4,29 4,64	1,54 1,58	7,99	6,5
	CD_3OD	4,55 4,61	1,54 1,47	4,78	6,6	XXXVI	CCl_4	4,48 4,61	1,60	11,98	6,8
	ДМСО-D_6	4,55 4,69	1,41 1,44	6,01	6,6	XXXVII	ДМСО-D_6	4,86 1,46	1,44	9,88	6,8
XXXIV	$CDCl_3$	4,41 4,74	1,56	7,95	6,4	XXXVIII	ДМСО-D_6	4,10 4,26	1,15	3,30 8,06	6,0

It is obvious that substitution of one of the hydrogen atoms bonded with nitrogen atom by the COH_3, COC_3H_7, COC_6H_5, and $SO_2C_6H_4CH_3$ groups leads to essential change of δ value of the NH proton signal. These data along with the values of signals integrated intensity allow to assume that the studied compounds exist (at least, predominantly) in one tautomeric form, namely amine one.

In this line, the spectrum of compound XXXVIII is of special importance. Shape of this spectrum, recorded in DMSO-D_6, as well as shape of the spectrum transformed by addition of water to the sample can be due to the tautomeric transformation described by the scheme

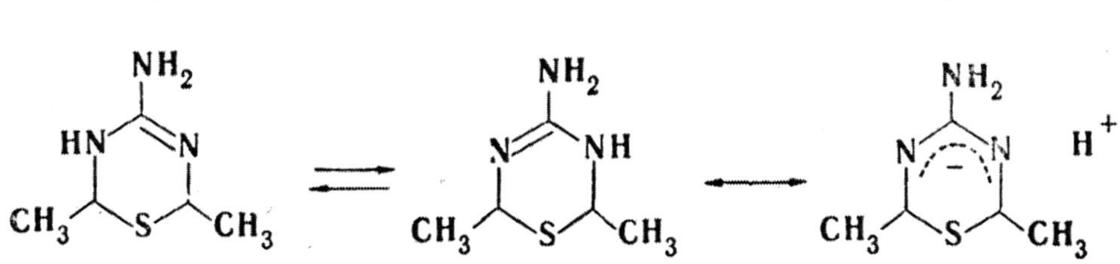

XXXVIIIa XXXVIIIb XXXVIIIc

When the mutual transformation of structures XXXVIIIa and XXXVIIIb is quite fast, N-3 and N-5 atoms become equivalent that is confirmed by structure XXXVIIIc. The latter can be stabilized via the formation of pseudo-allylic anion as shown below:

$$-^{-}N-C=N-\Leftrightarrow -N=C-N^{-}-$$

DMSO, forming strong intermolecular hydrogen bonds with labile protons, stabilizes compounds *a* and *b*, thus preventing fast exchange (in the NMR scale) of NH and NH_2 protons which is confirmed by nonequivalence of methine protons in this solvent. Addition of H_2O to the sample destroys this nonequivalence due to proton exchange with the medium.

Transformation of the PMR spectrum of compound XXXIII with paramagnetic additives also supports the amine form. Indeed, when $Eu(FOD)_3$is added to the sample, the signal of one of the methine protons is shifted by 2,3 higher that the signal of second proton. The shifts of the methyl group signals differ by the same value. The lower shift belongs to signals of the CH_3-group bonded with methine moiety, which signal has the lowest paramagnetic shift. This allows to assume that the observed paramagnetic shifts are caused mainly by coordination of $Eu(FOD)_3$ with the ring nitrogen atom in structure a. The conclusion on predominant existence of the studied compounds in amine proves that their most stable conformation is "semi-chair". Lack of the doubled signals in the spectra testifies that these compounds are isomerically similar (they have one variant of the CH_3-group position relative to the ring plane). It is difficult to make the unambiguous choice between *cis-* and *trans*-isomers. However, the Dreiding model of compound XXXIII, taking into account the observed paramagnetic shifts, gives preference to the *cis*-axial isomer.

2.4. Application of Paramagnetic Complexes for the NMR Spectra Analysis

The first publications devoted to application of paramagnetic complexes for the analysis of the NMR spectra bring forth a number of research problems. Among the latter, the most important are the following.

1. Extension of shifting and broadening reagents scope (first of all, LSR).
2. Search for the conditions of these reagents usage depending on the nature of functional groups, with which the substrate interacts with paramagnetic ions; study of solvent and temperature effect on character of LSR – substrate interaction; structural investigations of the applied paramagnetic additives; determination of the complex stoichiometry in solution, responsible for the induced paramagnetic shifts and broadenings.
3. Determination of efficiency of the paramagnetic complexes application for the analysis of NMR spectra of other than ^{1}H nuclei.
4. The solution of the theoretical issues concerning the application of the observed paramagnetic shifts and broadenings in structural studies.
5. Establishment of the nature of HFC in paramagnetic complexes with organic ligands (especially, this relates to lanthanides complexes).

Some of the above problems still represent a challenge. Below we will discuss selected publications dealing with the solution of these problems. Perhaps, some of them are interesting only from the historical viewpoint. However, these are pioneering works, which have laid the foundation of current advancements in the application of paramagnetic complexes for the spectral analysis.

As it has already been mentioned, after Hinckley has published his works, Sanders and Williams also report that $Eu(DPM)_3$ is an excellent shifting reagent [8]. Being coordinated with lone electron pair of the functional group, this europium compound induces considerable paramagnetic shifts. For example, in the PMR spectra of benzylic alcohol, aromatic protons are responsible for a broad singlet. After addition of $Eu(DPM)_3$ to a solution, the spectrum becomes available to the analysis according to the rules for spectra of the first order. The induced shifts quickly decrease with the increase in number of the bonds between hydroxyl group and the corresponding protons. When $Eu(DPM)_3$ is added to a solution of N-hexanol, spectrum of the latter can also be analyzed according to the rules for spectra of the first order (Figure 2.35).

Signal of the OH group broadens so considerably that the attempt to record it fails. However, the broadening of resonance lines of other protons is much lesser. It is found that $Eu(DPM)_3$ induces also paramagnetic shifts in the PMR spectra of ketones, ethers, esters, and amines. Ketones undergo complexing with 35 – 40% less efficiency than alcohols, while ethers and esters are even less effective ligands. However, amines behave like alcohols. Sanders and Williams have noted that $Eu(DPM)_3$ can be extremely effective for structural determination of natural compounds. For example, they have analyzed the PMR spectra of lupeol, recorded with additives of $Eu(DPM)_3$. The conditions of the latter reagent applicability have also been discussed [8]. In particular, it is underlined that it is well soluble CCl_4.

The efficiency of $Eu(DPM)_3$ application has been shown by Demarco et al. on the example of spectral analysis of *trans*- and *cis*-4-butylhexanol, isoborneol and borneol [205]. Signals in the spectra of these compounds have been assigned by comparison of paramagnetic shifts with the corresponding distances r' from resonating nuclei to the hydroxyl group oxygen (not to the coordination ion). Such approach is motivated by the fact that $(3\cos^2\theta - 1)$ values for protons of the studied compounds are different (see section 1.3 of this monograph). However, it is impossible to calculate these vales due to the lack of data on geometry of the paramagnetic complexes in solution. It has been established that the dependence between shifts and distances from protons to oxygen atoms in logarithmic scale is expressed by right line having the slope ratio of (– 2,3). This confirms that the induced shifts correspond to r':

$$lg(r')^{-3} = (-3)lg\,r'. \qquad (2.30)$$

The replacement of r by r' can be considered as justifiable, since it means shift of calculation start of the base number r for functional dependence $y = r^3$. It is obvious that for small values, the application of r for comparison of the paramagnetic shifts will give less satisfactory results. Therefore, the usage of r' instead of r can be most successful for the protons, which are most remote from europium atom. Figure 2.36 illustrates results of work [205]. In the PMR of spectrum of *cis*-4-*tert*-butylhexanol solution (Figure 2.36, curve 1) containing no $Eu(DPM)_3$, assignment of only two signals is possible. However, when the molar ration alcohol/Eu $(DPM)_3$ is 0,7, the recorded spectrum can be interpreted according to the rules for spectra of first order (Figure 2.36, curve 5). The signals have been assigned on the basis of values of spin-spin coupling constants using the double resonance experiments.

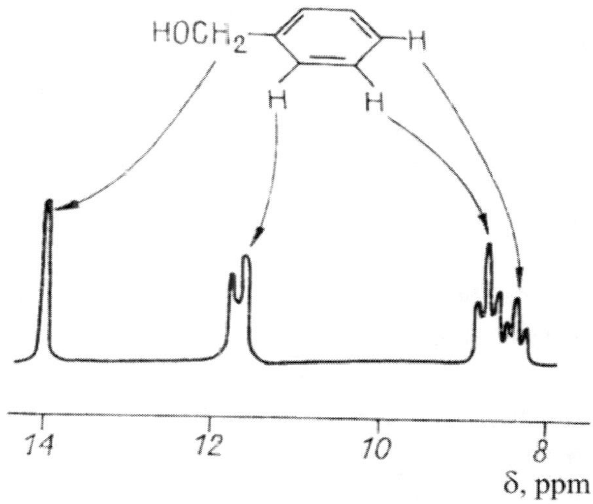

Figure 2.35. PMR spectrum of *n*-hexanol recorded with additive of $Eu(DPM)_3$

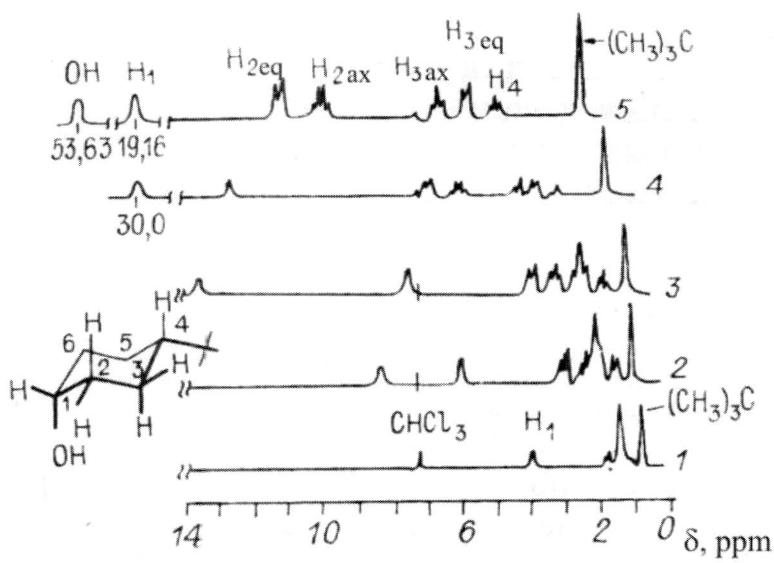

Figure 2.36. PMR spectra of *cis*-4-*tert*-butylhexenol without (1) and with (2-5) different additives of $Eu(DPM)_3$ (signal of methyl protons is recorded upon lower amplification as compared with other signals).

Demarco et al. have also illustrated efficiency of the shifting reagent method in structural studies of steroids and triterpenes. For example, the detail analysis of the PMR spectra of 5α-androstan-2β-ol and fridelan-3β-ol has been performed [206].

A possibility of the shifting reagent method application for determination of number and positions of the methyl groups in molecules of the natural compounds is confirmed by comparison of the shifts induced by $Eu(DPM)_3$ in the PMR spectra of twenty two triterpenes [207]. To record the spectra with non-overlapping signals, both concentration and temperature dependences of the observed shifts have been studied. The compared values of these shifts are standardized relative to the low-field signal of the 4-CH_3-group protons, the paramagnetic shift value of this signal being accepted as 10 ppm.

Briggs et al., [208] have reported for the first time that $Pr(DPM)_3$, like $Eu(DPM)_3$, induces paramagnetic shifts in the PMR spectra. It should be emphasized that these shifts are approximately three times as much than the shifts induced by $Eu(DPM)_3$ and have the opposite signs.

Metal	Paramagnetic shift ppm	Metal	Paramagnetic shift ppm
Ce	– 1,5	Dy	Not observed
Pr	– 3,0	Ho	– 17,0
Nd	– 1,75	Er	+ 4,4
Sm	– 0,27	Tm	+ 6,9
Eu	+ 1,33	Yb	+ 6,66
Gd	Not observed	Lu	0
Tb	– 19,2		

Values of the paramagnetic shifts induced by the complexes of $M(ClO_4)_3\cdot\{OP[N(CH_3)_2]_3\}_4$ type in the PMR spectrum of acetonitrile solution are given below. The shifts are given relative to the acetonitrile complex, when M is a lanthanide; the minus corresponds to the high-field shift [208].

The given example indicates that the complexes of other lanthanides, apart from $M(DPM)_3$, can be used as the shifting reagents. At the same time there are essential limitations in applications of these complexes. Some of them are listed below [208].

1. $M(DPM)_3$ form complexes with the coordination number 6, which are suitable for the interaction of organic molecules.
2. The shifting reagents method is based on the application of non-polar solvents.
3. The studied compounds cannot be coordinated with lanthanide reagents because of the steric hindrances; complex formation process may exert insignificant effect on the signals of resonating nuclei. For example, it has been found that $Pr(NO_3)_2(OPPh_3)_3$, $Pr(ClO_4)_3\{OP[N(CH_3)_2]_3\}_4$, $Eu(NO_3)_3(OAsPh_3)_4$, $Ce(NCS)_3(OPPh_3)_4$ and lanthanide perchlorates induce slight paramagnetic shifts (the signal of $\alpha-$protons is shifted only by approximately 0,5 ppm) in the spectra of alcohols and ketones. It should be noted that currently mainly $M(DPM)_3$ as well as non-polar solvents are used as LSR.

The observed and calculated paramagnetic shifts induced by $Pr(DPM)_3$ have been compared for the first time in [209]. A molecule of borneol XXXIX having the fixed conformation is used as a ligand. Unambiguousness in assignments of the observed paramagnetic shifts is reached by preliminary analysis of the compound XXXIX PMR spectrum recorded at three frequencies (60, 100 and 220 MHz) using the double resonance and deuteration.

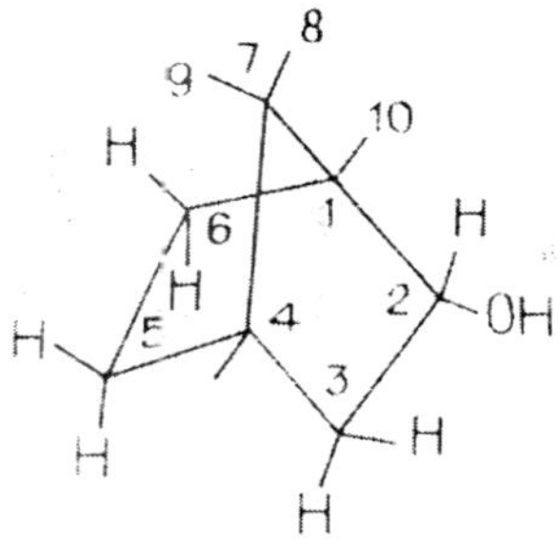

XXXIX

It has been established that: 1) the paramagnetic shifts are due to pseudo-contact interaction; 2) the complex in solution is axially symmetrical. This means that the induced shifts should be proportional to $(3\cos^2\theta-1)\cdot r^{-3}$. Orientation of the resonating nuclei is given by distance $\boldsymbol{Pr-O}$, angle $\boldsymbol{Pr-O-(C-2)}$ and dihedral angle $\boldsymbol{Pr-O-(C-2)-(H-2)}$. Further, the values of these parameters in the initial model (2,6 Å; 109° and 33°) are varied to achieve the agreement between the calculated and experimentally found shifts. Such an agreement is observed for the following values: 3 Å; 126° and 25°.

The conducted investigations allow concluding [209] that the problem of the hydroxyl group orientation in various complexes of rare-earth elements should likely be solved separately in each specific case. Importance of the $\boldsymbol{(3\cos^2\theta-1)}$ contribution to paramagnetic shifts (especially for the resonating nuclei located close to the coordinating ion) has also been revealed. This conclusion confirms the PMR spectrum of 3-tetracyclic epoxide XL, recorded in the presence of $Eu(DPM)_3$.

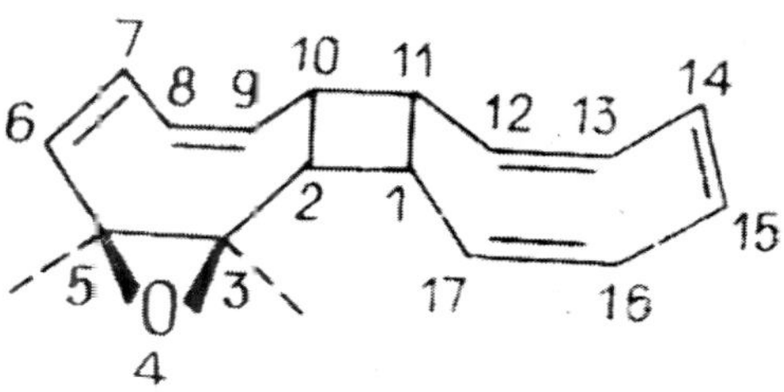

XL

It turns out that the signals of H-15 and H-16 protons (unlike the signals of other protons) are high-field shifted [210]. This is explained by the fact that $\theta_{i-15}=58°$, and $\theta_{i-16}=55°$, i.e. it is higher than 54,7° (if $\theta>$ 54,7°, $\boldsymbol{(3\cos^2\theta-1)}<0$). Therefore, the complex formation leads to additional shielding of H-15 and H-16 protons and shifting their signals towards the high-field. Other works also report on the high-field shifts of the signals induced by $Eu(DPM)_3$. It should be underlined that the authors of the publication [211] consider their assumption on contact and pseudo-contact mechanisms of coupling between unpaired electrons and resonating nuclei as the most rational explanation of the observed paramagnetic shifts towards the high-field. In their opinion [211], these two mechanisms should lead to appearance of paramagnetic shifts with opposite signs, as it is observed in the PMR spectra.

It is reported [212, 213] that ytterbium additives $Yb(DPM)_3$ are also used for the analysis of PMR spectra. The work [212] deals with primary, secondary and tertiary amines. It has been established that the value of induced paramagnetic shifts depends on the sterich factor as well as on nitrogen atom charge. The authors of this work have compared the values of paramagnetic shifts and $\boldsymbol{r}$ for pyridine and quinoline, which are characterized by the known orientation of the lone electron pair orbital. The distance N ··· Yb varies from 1 to 5 Å. It is found that when this distance is 3 ± 0,3 Å, graphic dependence of $\boldsymbol{\delta'}$ upon $\boldsymbol{r}$ (in logarithmic scale) is expressed by the right line. In the work [213], efficiency of $Yb(DPM)_3$ application for signals assignment in the PMR spectra of ketones and aldehydes is illustrated. Special works are dedicated to application of the shifting reagents method to the compounds having more than one coordination center. For example, it appears that *trans*- and *cis*-1-hexyl-2-phenylacetidine-3-ol is coordinated with $Eu(DPM)_3$ by the hydroxyl group oxygen, but not by nitrogen atom of the four-membered cycle [214]. The detail analysis of the paramagnetic shifts in the spectra of compounds having several coordination centers is given in the works [215,216].

Dependence of the pseudo-contact coupling not only upon $\boldsymbol{r}_i$, but also upon $\boldsymbol{\theta}_i$ complicates assignment of the signals by comparison of the observed paramagnetic shifts. However, in the case of fast exchange, paramagnetic broadenings of the resonance lines at given temperature of the sample depend only upon $\boldsymbol{r}_i^6$. Therefore, if the signals are shifted, their broadening can be determined, but $\boldsymbol{\Delta}$ values should be used for the same purposes as $\boldsymbol{\delta'}$. This has been taken into account in the work [217], where it is disclosed that not only shifts, but also the broadenings of the resonance lines induced by ions of the rare-earth elements, allows to use them as excellent NMR-probes for the investigations of biological systems. In particular, it has been illustrated by the analysis of lysozyme PMR spectrum recorded in the presence of $\boldsymbol{Eu^{3+}}, \boldsymbol{Ho^{3+}}, \boldsymbol{Gd^{3+}}$ ions. The distances from the coordination ion to proton of acetamide and glycoside methyl groups have been calculated by comparison of signals broadenings induced by gadolinium.

A comparative assessment of some $M(DPM)_3$ complexes (M = Pr, Sm, Eu, Tb, Ho, Yb) has been carried out [218]. Among the criteria of such comparison are: (i) the positions of proton signals of DPM *tert*-butyl groups; right line gradients of the $\delta' = \delta'(P)$ dependence, where $P-$is a molar ratio $M(DPM)_3$/studied compound; relative shifts in the complexes of $M\{OP[N(CH_3)_2]_3\}_4$ type; the ratio broadening/proton signal shifts of the *tert*-butanol methyl groupsin the presence of $M(DPM)_3$. The comparison has shown that $Eu(DPM)_3$ turns out to be best additive for recording the PMR spectra of the first order with preservation of multiplet structure. The signal of *tert*-butyl groups of this LSR is observed in higher field as compared with the signal of tetramethylsilane, thus not overlapping with the signals of the studied compounds. Besides, $Eu(DPM)_3$-induced paramagnetic shifts towards low-field are more suitable from the experimental point of view. $Pr(DPM)_3$ should be used in addition to $Eu(DPM)_3$. One should remember that the signal of $Pr(DPM)_3$*tert*-butyl groups, appearing in the region 3 – 5 ppm, can be overlapped with the signals of the studied compounds. The detail analysis of $M(DPM)_3$ application as shifting reagents as well as possibilities of the LSR method is provided in the work [219].

The conducted investigations have shown that lanthanide-induced paramagnetic shifts significantly depend on structure of diketone. It is found that europium *tris*[3-*tert*-butylhydroxymethylene-(+)-camphorate] can be employed for determination of enanthiomeric [220]. However, application of this compound turns out to be limited. Europium *tris*[3-heptafluoropropylhydroxymethylene-(+)-camphorate] [221] appears to be more efficient optically active shifting reagent. At the same time, the application of $Eu(DPM)_3$ for these purposes can also be successful [222]. Already in the early seventies the last century, there were publications concerning the NMR studies involving other (apart from protons) nuclei. In these studies, additives of rare-earth element complexes were used. The effects of $Eu(DPM)_3$ and $Yb(DPM)_3$ additives on the ^{14}N NMR spectra of pyrrolidone, *n*-propylamine, pyridine, N-methylimidazole, acetonitrile, dimethylformamide [223] have been investigated. Briggs et al. have shown the efficiency of the ^{13}C NMR spectra interpretation with a help of *tris*(dipivaloylmethane)lanthanides [224].

A method for assignments of signals in the ^{13}C NMR spectra has been developed. The method comprises the combined application of $M(DPM)_3$ and heteronuclear double resonance $^{13}C-\{^1H\}$ [225]. It is also reported on the nature of lanthanide-induced shifts in the ^{31}P NMR spectra [226]. In particular, it is found that $Eu(NO_3)_2{\cdot}6H_2O$ is an effective shifting reagent in polar solvents for compounds bearing the $X-O^-$ $(X=R-CO-, R_3\overset{+}{P}, R_3\overset{+}{As})$ functional group. Interesting results on isotope effect of deuterium have been obtained [227,228]. We also should mention the works [229,230], where the statistical method for the analysis of paramagnetic shifts induced by rare-earth element complexes is discussed. Finally, the review [231] is devoted to application of the shifting reagents and contains complete list of references published in 1969 – 1971, and a part of works published in 1972.

The works which have appeared after publications of papers [7, 8], define a scope of applicability of the LSR method. Besides, they pave the way for application of the rare-earth element complexes in the "active" purposes (see section 1.2.2 of this monograph). The authors of the work [232] employ $Yb(DPM)_3$ additive for determination of conformation and configuration of 1,6-dibromo-3,7-dimethyl-3,4-epoxyketone XLI. The PMR spectrum of compound, recorded without additive of the shifting reagent, shows that according to ***J*** constant value, this ketone has the *trans*-orientation of protons at the double bond. From the same spectrum (on the basis of experiments on the Overhauser intranuclear effect) follows that H and CH_3-groups have *cis*-configuration of N and CH3-group relative to the epoxy ring. However, more detail analysis of stereochemistry turns out to be impossible. The comparison of the experimental and calculated paramagnetic shifts (performed taking into account the rotation around the C-3 and C-4 bonds) has confirmed the predominant relative disposition of H-5, H-5′ and H-6 protons. For further studying of XLI configuration, the paramagnetic shifts with their averaging on three rotamer in each of two possible positions of the $CCl(CH_3)_2$-group have been calculated.

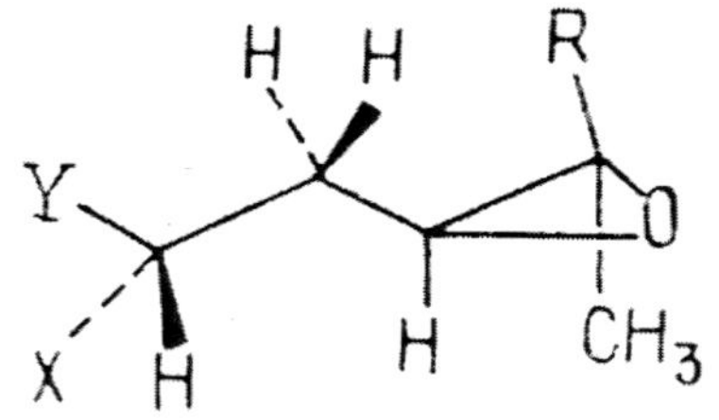

XLI

X= – CCl$(CH_3)_2$;Y = Br; R = – CH=CHBr.

Agreement factor ***R*** between the calculated and experimental ω values is accepted as criterion of the calculations, which are performed using special computer program [229,230]:

$$\boldsymbol{R} = \left[\frac{\sum v_i \left| \omega_{i\,exp} - \omega_{i\,calcd} \right|^2}{\sum v_i \left(\omega_{i\,exp} \right)} \right]^{1/2}. \qquad (2.31)$$

Here v_i characterizes a probability $\omega_{i\,exp}$ values determination. The $\omega_{i\,calcd}$ values are calculated by the MacConnel – Robertson equation. It is found that the calculated and experimental shifts coincide, if the (CH_3)-groups have *heme*-orientation at the C-6 atom.

Eu$(DPM)_3$ is also employed for determination of XLII, XLIIIa-c oximes configuration [233].

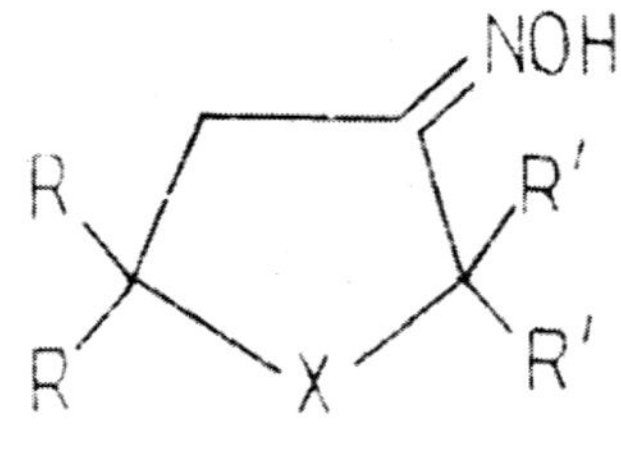

X=CH_2; R=R′=CH_3

X=O; a: R=R′=CH_3; б: R=CH_3, R′+R′=$(CH_2)_5$;

в: R+R=$(CH_2)_5$, R′=CH_3.

The values ω_{exp} and ω_{calcd} are correlated. The origin of the coordinates is combined with the coordinating atom (N or O) so that its bond with the closest atom of other fragment of the molecule is located on the positive direction of axis *x*.

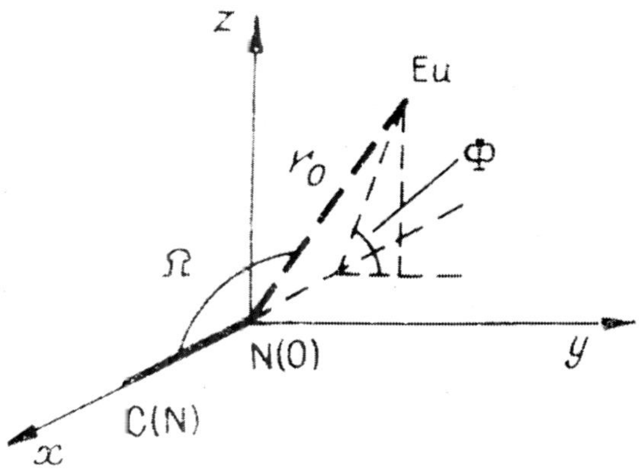

In this case, the position of Eu is defined by three parameters: distance to coordination atom (r_o) and Ω and Φ angles. The relative position of Eu is determined by varying the r_o value within 2 – 4 Å, the angle Ω changes from 100° to 240° (coordination at the nitrogen atom) and from 30° to 180° (coordination at the oxygen atom). It is assumed that (i) the five-membered oxime cycle is planar;(II) the europium atom is located in the plane of this cycle, i.e. Φ = 0 or 180°.The calculations performed allow to infer that (taking into account the orbital orientation of oxygen and nitrogen lone pairs) the value Ω related to r_{0min} corresponds to the *Z*-configuration of compound XLII. Analogously, it has been concluded that three other studied compounds are also of the *Z*-configuration.

The authors of the work [234] have employed LSR for the first time in ^{19}F NMR stereochemical studies. Positions of two atoms in the benzene cycle are determined and the substitution stereochemistry in the five-membered trifluoroindoline cycle XLIV is investigated.

^{19}F NMR spectra recorded without LSR additive of LSR have confirmed that compound XLIV contain one fluorine atom in the position 2 and two fluorine atoms in the aromatic ring. However, it is impossible to determine the positions of these atoms in the ring as well as relative configuration of F-2 and the hydroxyl group. Comparison of the experimental paramagnetic shifts with calculated ones (for possible structures) allows authors of the work [234] to conclude that fluorine atoms are located in the positions 6 and 7 of the benzene ring, and F-2 and OH- group have relative *cis*-orientation.

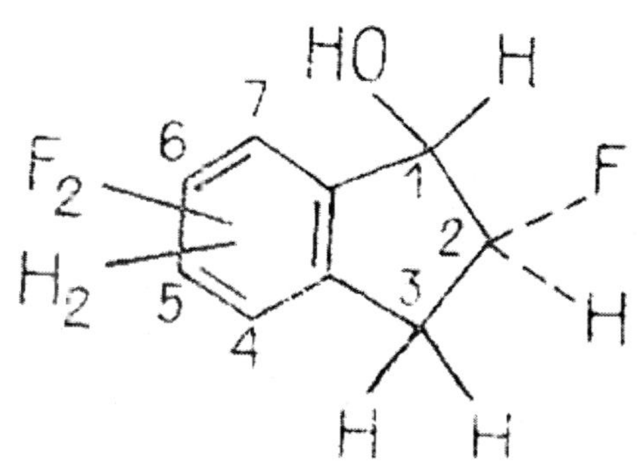

XLIV

The solutions of conformational problems with a help of LSR are discussed in the paper [235].

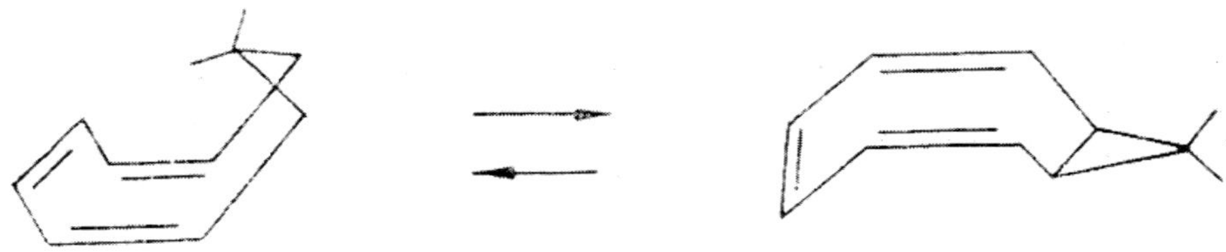

XLVa XLVb

The derivatives of bicyclo-[6,1,0]-none-2,4,6-triene XLV can have folded (XLVa) and extended (XLVb) conformations, which can exist in equilibrium in the solution. The authors of the above work have compared the experimental shifts with calculated **ω.** values (using McConnell – Robertson equation). The calculations were performed in the following conditions. First, the europium atom is localized relative to H-1, H-8, H-9 and H_X protons. Then paramagnetic shifts **ω** of other proton signals for each conformation are calculated. It is assumed that europium atom is located in the symmetry plane of the ligand molecule. Comparison of the experimental and theoretical values involves determination of R_{min} (see expression 2.31). Eventually, it has been found that the studied compounds are of the extended conformation. If conformation of XLVa is realized, its contribution does not exceed 5%.

The limitations of approaches to determination of structures of acyclic molecules in solution have motivated the authors of the work [236] to check applicability of LSR in such studies. For this purpose, the conformational analysis of all possible eighteen aliphatic alcohols having six carbon atoms has been carried out. In the study, paramagnetic shifts in ^{1}H and ^{13}C NMR spectra are analyzed. Geometrical factors $\varphi = \left(3\cos^2\theta - 1\right)/r^3$ from McConnell - Robertson formula and ω_{exp} values are compared. Using the Dreiding models [236], the relative location of the resonating nuclei (^{1}H, ^{13}C) and lanthanide is selected so that the correlation would not be less than 0,950, and the plot of ω_{exp} dependence upon φ would represent (in logarithmic scale) a right line with the tangent of angle equaling to – 3. The distance between the lanthanide and alcohol molecule oxygen is accepted to be 3,0 Å, while C— O —Ln angle = 130º. Analysis of the experimental data thus performed makes it possible to establish preferable average conformations of all studied compounds, the conformations determined by ^{1}H and ^{13}C NMR spectra being coincided. The conformations determined have been discussed [236], and where it is possible, the conformer prevailing in the complex is specified. Analysis of the J values enables to conclude that in some cases, the application of LSR shifts equilibrium between conformations towards one of them. For example, in 3,3-dimethyl-1-butanol, upon coordination of the alcohol with paramagnetic additive, the equilibrium between the *trans*-and *gauche*-conformations is shifted towards the trans-conformer. The information on spatial structure of the studied alcohols obtained by the authors of the discussed paper turns out to be quite comprehensive and detailed in order to understand the relationship between molecular conformation and parameters of [13] NMR spectra and to discuss a problem of stereoisomers identification using the NMR spectra.

A structural study of ionic pairs formed by cations of N-alkyl-N-benzylpyridine series with lanthanides ($Ln = Tb, Er, Yb, Tm$) has been performed [237].

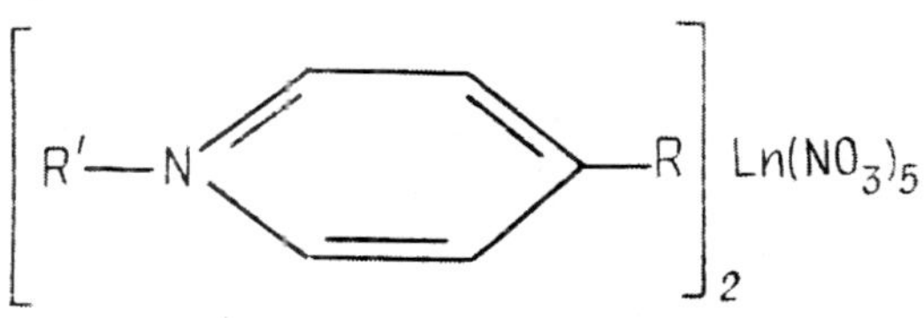

XLVI $R = CH_3$; $R' = 2,4,6\text{-}(CH_3)_3C_6H_2CH_2$

XLVII $R = C_6H_5$;

$R' = CH_2 - CH_2 - CH2 - CH_2 - CH_2 - CH_2 - CH_2 - CH_3$

1 2 3 4 5 6 7 8

Structure of the complex in solution is determined by comparison of the experimental shifts with calculated ones. The absence of dependence of relative paramagnetic shifts upon nature of the lanthanide is considered as the proof of formation of similar structural complexes in solution. Positions of the anions are defined by r_0, Ω, Φ parameters (Scheme XLVIII). It is suggested that the main magnetic axis should not necessarily pass through the nitrogen atom.

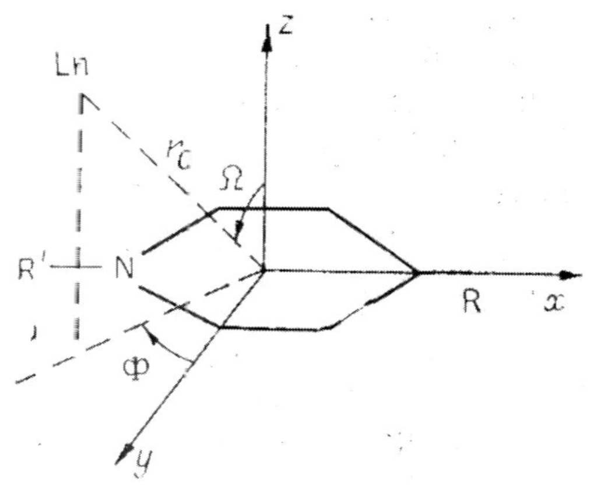

XLVIII

Therefore, parameter x_0, defined as distance between nitrogen and the origin of coordinates, is used in the calculations. Two models are considered: 1) static one where the relative position of anion and cation is conditioned only by a set of the fixed values $\boldsymbol{r_0, \Omega, \Phi, x_0}$; 2) geometry of the complex is defined by a set of the same parameters, but calculated when fast rotation of the cation around axis x takes place. A model of free rotation provides the best agreement between compared values for signals of the pyridine and benzene ring protons. The effective cation radius (~ 3 Å) ensures free rotation of the pyridine ring [237] and indicates the absence of solvation layer of the solvent molecules between cation and anion. The free rotation model provides also good agreement of the calculated and experimental values for H-5 ÷ H-8 protons of the octyl chain, but not for H-2 ÷ H-4. This model allows explaining the shifts, which values are significantly higher than the experimental ones. This is obviously due to the fact that H-2 ÷ H-4 protons are located closely to anion in the course of rotation. In its turn, the static model gives better results for H-2 ÷ H-4, than for H-5 ÷ H-8. Therefore, it has been concluded [237] that the remote fragments of the octyl chain rotate freely, while the fragments located near to the coordination center have predominant orientation, trying to avoid contact with the anion.

The development of the paramagnetic additives method expands possibilities of the NMR spectroscopy in conformational studies of biologically important molecules. However, in some cases, the application of this method leads to the results, which do not completely agree with the data obtained from the analysis of spin-spin coupling constants. Therefore, in opinion of the authors of the work [238], it is important to check comparability of two methods by using suitable compounds. In this line, conformational study [238] of 3′, 5′-cyclic trimidine phosphate XLIX in aqueous solution has been performed. PMR spectra of this compound are recorded with additives of paramagnetic ions $Pr^{3+}, Eu^{3+}, Nd^{3+}, Yb^{3+}$. Unfortunately, the spectra recorded with additives of the latter ions turn to be inappropriate due to strong broadening and (or) overlapping of the signals.

It is found that relative paramagnetic shifts (normalized with respect to shift of the H′-3 proton signal, which $\omega =$ 1,00), induced by Pr^{3+} and Eu^{3+}, are satisfactory. Therefore, it is concluded that these shifts are pseudo-contact, and the magnetic tensor is axially symmetrical, i.e. McConnell – Robertson formula can be used for the calculation. The values of J_{HH} and J_{PH} constants do not almost change after addition of lanthanides to the sample. Thus, conformation of molecule XLIX remains intact during the coordination. Also, PMR spectra of compound XLIX have been recorded [238] with Gd^{3+} additives and the corresponding values of relaxation times T_1 have been determined. The found $(T_1)_i$ values, normalized with respect to T_1 of H′-3 proton, are also used in the stereochemical analysis. The known compound 3′,5′-cyclic uridine phosphate is used as a standard, which then is transformed to structure XLIX by iterative procedure.

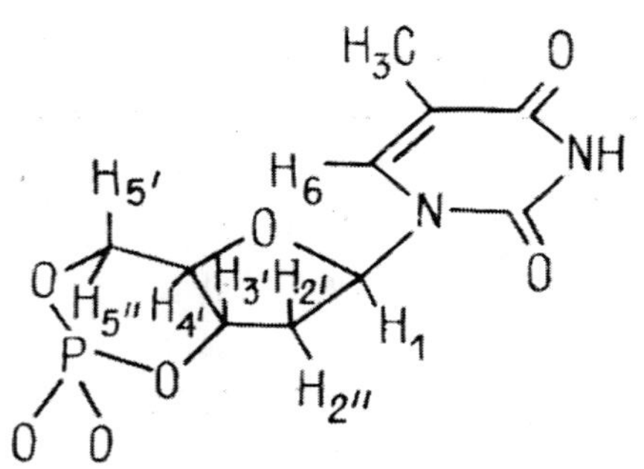

XLIX

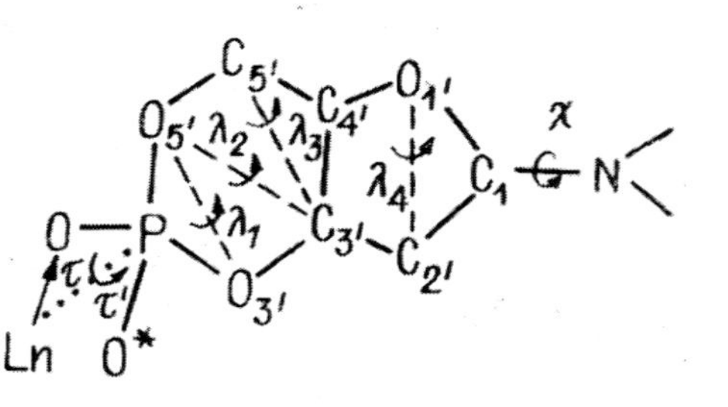

L

Among the varied parameters are $\lambda_1 - \lambda_2$. They are defined as dihedral angles between the planes formed by three atoms. For example, λ_1 characterizes the angle between $PO_{5'}$, $O_{3'}$ and $O_{5'}$, $O_{3'}$, $C_{3'}$ (L) planes. It is assumed that the lanthanide is located on bisector of the ORO* angle. The distance Ln – P is accepted to be 3,0 Å. The minimum value of the $\boldsymbol{R}$ factor is calculated. The $\boldsymbol{R}$ factor is defined by the following expression:

$$R = \left[\frac{\sum v_i \left(\omega_{i\,exp} - \omega_{i\,calcd}\right)^2 + \sum u_i \left(T^{-1}_{1\,i\,exp} - T^{-1}_{1\,i\,calcd}\right)^2}{\sum v_i \omega^2_{i\,exp} + \sum u_i \left(T^{-1}_{1\,i\,exp}\right)^2} \right]^{1/2}, \tag{2.32}$$

where u_i have the same sense as v_i (see expression 2.31). By factor $\boldsymbol{R}$ minimization via variation of $\lambda_1 - \lambda_4$, χ, τ, τ' angles, it has been established that the phosphate ring of compound XLIX has chair conformation. This conclusion is in agreement the expected data obtained from analysis of vicinal constants J_{HH} and J_{PH}. Owing to removal of H-1, H-6 protons and CH_3-group from the coordination center, the $\boldsymbol{R}$ value is insufficiently sensitive to alteration of λ_4 that complicates determination of the dioxaribose ring conformation. Therefore, the authors of the discussed work have optimized $\boldsymbol{R}$ factor by varying the $J_{12'}$ and $J_{12''}$ constants. This gives a chance to carry out the conformational analysis of the whole molecule of compound XLIX.

The examples discussed here and in the section 2.3 of this monograph outline the main circle of problems which can face the researchers, when using additives of paramagnetic complexes for study of molecular structure in liquid phase. The same examples illustrate the tools, which need to be used for the purpose of obtaining structural information from the spectra of nuclear magnetic resonance transformed by complex formation process.

In conclusion of this section, it is pertinent to cite the reviews [5, 107, 170, 171, 158, 239 – 248], which cover the works related to application of paramagnetic complexes in the NMR spectroscopy.

2.4.1.Experimental Procedures

The paramagnetic additives method comprises the recording of NMR spectra with or without additives of a paramagnetic complex. Molecules of solvent should not coordinate with paramagnetic additive. This is the main requirements which are imposed to solvents. When the lanthanide shifting reagents are employed, the typical solvents are CCl_4, $CHCl_3$, and benzene. Quite often CH_3OH is used as a solvent, if compounds, coordinated with lanthanide shifting reagents by pyridine nitrogen atom, act as a substrate. Deuterated analogs of these solvents now became readily available. If elements of the iron group are employed as paramagnetic additives, the suitable solvents are (apart from $CHCl_3$) CH_3NO_2 and DMSO. HMDS is used as internal standard.

Usual temperature for recording of the NMR spectra is 45 ÷ 50° C, if LSR is employed as an additive. For iron additives, the temperature can be significantly higher: everything depends on realization of the fast exchange conditions. Identical broadenings of all signals in the NMR spectra irrespective of the distance between resonating nucleus and coordination ion (see figure 2.6 present monographs) are strong indications of the absence of the above conditions. Concentration P_M of paramagnetic additive relative to the substrate dissolved in the sample should be $10^{-3} - 10^{-2}$. In this range, one should prepare 7 – 10 additives. This is sufficient, first, to be convinced in linear character of $\delta' = \delta'(P_M)$ dependence of the observed shifts, and second, to approximate (by least square method), if it is necessary, the above dependence in order to determine paramagnetic shifts ω in the complex.

As our experience shows, to obtain qualitative ^1H spectra, it is sufficient to dissolve about 10 mg of substrate (with molecular weight of 100 - 200) in a sample having standard volume (0,7 cm^3). Such amount of substance is sufficient to quickly record qualitative spectra (with a good signal/noise ratio). Certainly, the modern spectrometers, operating in the pulse mode, allow to work (in principle) almost with any quantity of the substrate. Here we mean concentrations, which make it possible to obtain the relevant spectral information during small periods of time. The spectra in Figure 2.37 show a real ratio between content of the substrate in the studied sample and paramagnetic additive. The first spectrum belongs to the sample, which solution (~ 1 cm^3 in DMSO) contains 32,2 mg of EIM.

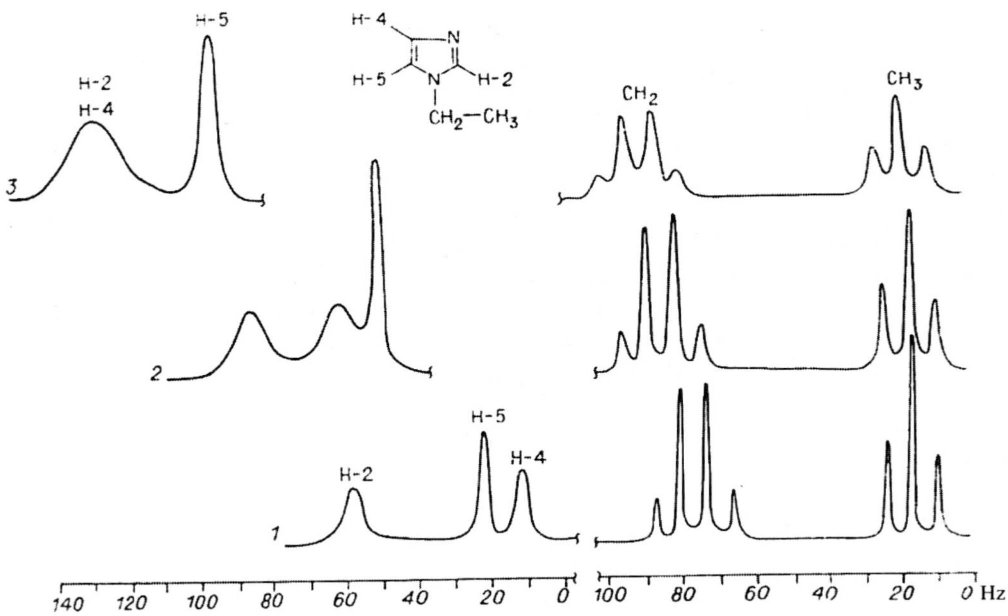

Figure 2.37. PMR spectra of EIM in DMSO (1) at 90° C with different concentration of paramagnetic additive $CoCl_2 \cdot 4EIM$ (mg): 0,3 (2); 1,9 (3).

From the viewpoint of the method application in structural studies, ^1H nucleus seems to be most appropriate, since HFC with participation of this nucleus can be interpreted with highest accuracy. It is also important that the nature contains 100% of ^1H isotope. Therefore, the PMR spectra can be recorded at insignificant concentration of the studied compound and paramagnetic additive, thus having minimized or even excluded the intermolecular effects on the observed paramagnetic shifts and broadenings.

Among the most widespread paramagnetic additives are $Eu(DMP)_3$,$Eu(FOD)_3$ (Ln – Eu, Yb, Pr, Gd);$M(AA)_2$ (M – Ni, Co, Mn); $MX_2{\cdot}mR$ (M – Ni, Co; X – halogen;R – molecule of the substrate which spectrum is analyzed).

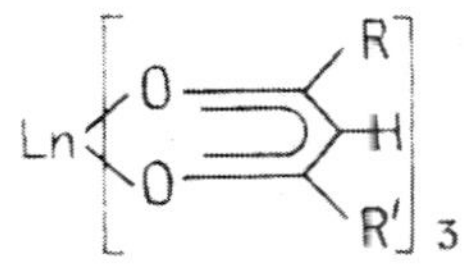

$R = R' = t\text{-}Bu \rightarrow Ln(DPM)_3$

$R = t\text{-}Bu;\ R' = C_3F_7 \rightarrow Ln(FOD)_3$

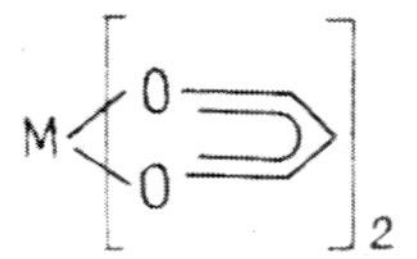

$M((AA)_2$

Molecular mass:$(DPM)_3 \rightarrow$ 549,813;$(FOD)_3 \rightarrow$ 885,838;$(AA)_2 \rightarrow$ 198,218.

As for the above-discussed range of paramagnetic additive concentration in the studied sample, we should give here the additional explanation, since there is an impression that (as it follows from a number of publications) the choice of $\mathbf{P}_{\mathbf{M}}$ value is problematic when it should not be. It is known that LSR form self-associates depending on the experimental conditions (concentration, temperature, solvent nature) (see [170] and the references cited therein). The spectrum shown is Figure 2.38 is very illustrative. It contains about 40 signals, which coalesce in two broad maxima ($\boldsymbol{t}-$Bu and CH resonances) at the temperature increase. The given example does not demand additional comments: one should strike for minimum of LSR concentration in the sample. Under the standard conditions of recording the NMR spectra transformed by paramagnetic additives, concentration of the substratum is centimole, while that of lanthanide is, at least, by 10 times lesser. In these conditions, one can suggest that the complex formation of LSR with the substrate should be a predominant process (instead of formation of self-associates).

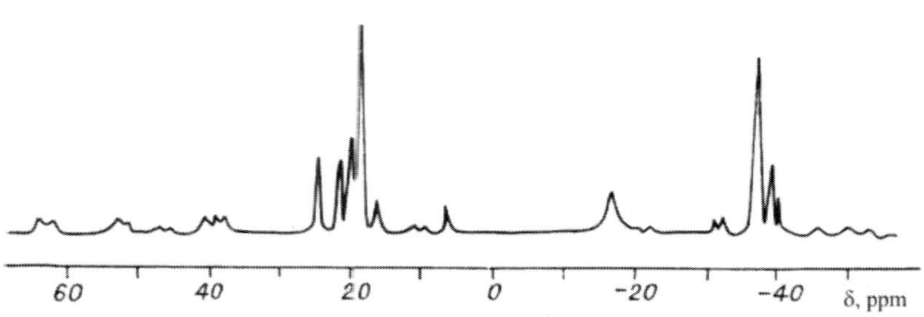

Figure 2.38. PMR spectrum of 0,1 M $Yb(FOD)_3$ in $CDCl_3$ at – 38° C.

In section 1.2.2, it has been mentioned that some researchers associate shifts $\boldsymbol{\omega}$ in the complex with necessity of spectra recording upon varying shifting reagent/substrate ratio in high range, including when $\mathbf{P}_{\mathbf{M}} > 1$. If to agree with such statement, the following question arises: complex of which structure is responsible for the paramagnetic shifts used in structural studies. In the review [170], it is stated that this problem can be solved by such recording of the NMR spectra, when signals from free and complexed substrate are detected. However, using the standard solvents, it is not always possible to reach such low temperatures, at which such detecting can be made.

In principle, one can use any $\mathbf{P}_{\mathbf{M}}$ value: everything depends on what aims are pursued. It is important to understand (this circumstance is not considered in a number of works), what approximations of the HFC theory are used for calculations of the experimental results. The section 2.1 provides a detail discussion of the relaxation processes and chemical exchange effects upon values of the observed paramagnetic shifts and signal broadenings. Undoubtedly, the so-called fast exchange, which conditions are defined by the expression 2.33, is preferable [158]

$$\mathbf{P}_{\mathbf{M}} << 1. \qquad (2.33)$$

However, just such $\mathbf{P}_{\mathbf{M}}$ values ensure the highest unambiguity of the structure of LSR/substrate complex, which is responsible for the induced shifts and broadenings. One should bear in mind that emergence of average spectrum does not mean automatic application of the expressions (2.7) – (2.23). Especially this relates to signal broadenings.

Electron states of Gd^{3+} and Mn^{3+} paramagnetic ions are characterized by high degree of symmetry [249]. Therefore, relaxation time of uncompensated electron spin in the complexes formed by these ions is rather high (as compared with relaxation times in complexes of other elements of iron and lanthanum groups, respectively). For this reason, HFC leads to drastic decrease of relaxation time of ligands nuclear spins and, as consequence, to strong broadening of the NMR signals. It is not surprising that the addition of very small amount of paramagnetic complexes, for example, gadolinium ($\mathbf{P}_{\mathbf{M}} \approx 10^{-4}$ and even lower) to the studied sample results in noticeable transformation of signal width without visible shifts. Such peculiarity of the spectrum provides a basis for the "broadening reagents" concept. Moreover, the long time of electron relaxation confirms strong dependence of relaxation time of nuclear spin and, hence, width of its NMR signals, upon $\boldsymbol{r}$ distance (see expressions (2.4), (2.5)). This phenomenon is observed, if the nucleus, which signal is broadened, does not participate in the contact HFC with uncompensated electron spin. Thus, it can appear that for a nucleus, conditions of slow exchange or the intermediate case, requiring the expressions (2.16), (2.17) for processing the experimental data, are realized.

For the quantitative analysis of paramagnetic shifts and broadenings, the need for application of highly pure LSR is obvious. However, it has been found that commercially available compounds $Ln(FOD)_3$ may contain considerable admixtures of binuclear complex $Ln(FOD)_3{\cdot}M(FOD)$ (M is an ion of alkaline metal), which cannot be removed by sublimation or recrystallization [170]. It is recommended to purify unclean shifting reagents (dissolved in CCl_4) with aqueous solution of the corresponding lanthanide salt or by treatment with 0,1 N HCl solution. Synthesis of lanthanide trisdiketonates is well developed (see [250] and the references cited therein).

Discussing the experimental conditions, the authors of this monograph mean, first of all, non-aqueous media which are the most common for recording the NMR spectra (including with paramagnetic additives).Specificity of aqueous media is well known. Naturally, owing to this circumstance, the application of paramagnetic complexes in aqueous media becomes a special problem (from experimental point of view, but not from positions of the HFC theory and chemical exchange) [170, 251).

It is important to remember that reactivity of heteroatom compounds relative to LSR has the following order [178, 244]: amines > hydroxyls > ketones > aldehydes > ethers > nitriles. A similar order for ability of LSRs to induce paramagnetic shifts has been also established (at the smallest broadening of resonance signals from substrate nuclei) [252]:

$$Eu >> Pr > Sm >> Nd \geq Yb \geq Tm, Dy, Tr \geq Er, Ho.$$

As it has already been mentioned in section 2.1, structural problems, which are covered in the present monograph, do not limit the application field of paramagnetic complexes additive. It is pertinent to add that usage of paramagnetic ions (first of all, Cr^{3+}) for facilitation of signals detection from other nuclei (apart from ^{1}H) is a rather routine procedure in the NMR spectroscopy. It is found that even substrates, containing no functional groups capable of coordination with the paramagnetic center, can for some time be predominantly oriented relative to the paramagnetic ion (because of electrostatic interactions, for example). This leads to decrease of relaxation time of the substrate nuclei. Mechanisms of interaction between organic molecules and paramagnetic relaxation reagents as well as peculiarities of ^{13}C and ^{15}N signals detection are discussed in the works [253, 254].

The effect of paramagnetic relaxation reagents is a special case of manifestation of magnetic interactions between electrons and nuclei. The character of such interactions causing diverse relaxation effects and, consequently, broadenings of lines in the spectra, is defined by the expressions (2.4), (2.5). Hence it follows that the decrease in relaxation time of electron spin leads to narrowing of the NMR signals (conditions of the NMR spectra recording improve) and vice versa. Such interrelation of relaxation times of electronic and nuclear spins allows one to influence the shape of the NMR and ESR spectra of paramagnetic systems (when electron-nuclear interaction takes place). For example, Cu(hexafluoroacetylacetonato)$_2$ and Cu(trofluoroacetylacetonato)$_2$, upon dissolving in chloroform (10^{-3} M), give well-resolved EPR spectra, in which the hyperfine structure is observed. When only 10^{-2} M of stable radical of di-*tert*-butylnitroxide is added to a solution, the ESR signals broaden so much that the hyperfine structure becomes forbidden, and it is impossible to record ^{19}F NMR spectrum of these chelates in chloroform. However, upon dissolving these chelates in the mentioned solvent, the recording of their ^{19}F NMR spectra becomes possible. This result is explained by the fact that the spin exchange between free radical and copper complex reduces the electron relaxation time, thus allowing the NMR signal to be detected [42, 255]. In principle, the mentioned dependence of nuclear spins and electrons makes it possible to investigate paramagnetic complexes on a basis of combined application of NMR and ESR techniques. For example, this approach has been used [151] for study of dopamine complexes with Cu^{2+}, Mn^{2+} and Fe^{3+} in aqueous solution.

Successful application of the paramagnetic additives method requires certain experimental skills. This is connected with some difficulties arising at the initial stage of research. The matter is that it is necessary to find such ratio of the substrate and paramagnetic additive concentration that the contents of the latter in sample would be insignificant. At the same time, sampling of this additive should be taken with sufficient accuracy. One should bear in mind that exact ratio between concentration of substrate and paramagnetic complex is important, when absolute values of paramagnetic shifts and broadenings are used in the structural studies. If we deal with only simplification of the NMR spectrum shape or with the quantitative analysis, relative (normalized) δ' and Δ values are employed. In these cases, the above requirements are, of course, not mandatory. Here it should be underlined that the clear understanding of the method applicability in structural studies and scope of HFC theory approximations defines informational content of the analysis of NMR spectra transformed by paramagnetic complexes.

2.4.2. Experimental Data Analysis

To obtain adequate information on the molecular structure, one should take into account all the factors influencing the induced paramagnetic shift in the NMR spectra. According to HFC theory (see section 1.3) the shifts caused by lanthanide shifting reagents can be generally presented by the following expression:

$$\omega_{\text{theor}}^{ijk} = D_j \frac{3\cos^2\theta_{ik} \quad 1}{r_{ik}^3} + D'_j \frac{\sin^2\theta_{ik}\cos 2\Omega_{ik}}{r_{ik}^3} + F_j A_{ik} \qquad (2.34)$$

where D_j and D'_j – are also constants, which depend upon nature of the lanthanide and defined by tensor of its magnetic susceptibility; $F_j = g_J(g_J - 1)J(J+1)$ is also defined by nature of the lanthanide and electronic state of the ion, J – is a value of total angular momentum of the ion; A_{ik} – is a constant of contact HFC.

Generally, among the unknown parameters are:

1) D_j and D'_j for each lanthanide;
2) three variables ($x_{ik}; y_{ik}; z_{ik}$) defining the position of resonating nucleus relative to lanthanide ion in the Cartesian system of coordinates;
3) A_{ik} for each nucleus. Thus, six parameters are unknown. Two of them (D_j and D'_j) relate to j – lanthanide, and four other ($x_i; y_i; z_i; A_i$) associate with i – nucleus.

If there are *n* nonequivalent nuclei and *m* lanthanides, the system of the equations (2.34) will contain the unknown (4*n* + 2*m*) and its strict solution requires the following condition

$$n \cdot m \geq 2m + 4n. \qquad (2.35)$$

For a number of reasons, analytical solution of the equation (2.34) is unreal. In practice, similar systems of the equations are solved iterative procedures. For this purpose, the criteria, which should reach any minimum value at coincidence of the experimental and calculated shifts, are applied. Details of such approach to lanthanide shifting reagents are given in numerous publications [229, 230, 256 – 259].In the USSR, this problem has been studied in detail by Kornilov et al., [260].

It has been reported [261] on development of subroutine package "LISMIN" intended for the solution of problems related to search for global minimum of many variables function. The package allows one to minimize the functions by optimizing the values of the corresponding parameters. Maximum number of such parameters is 150. At the same time, 55 parameters could simultaneously change. In the above package, two criteria are offered: the known $\boldsymbol{R}$ [229], and $\boldsymbol{R}^*$ set by the expression:

$$R^{\bullet} = \frac{1}{N} \sum W_{ijk} \left(\omega_{\exp}^{ijk} - \omega_{theor}\right)^2, \qquad (2.36)$$

where i – is a number of nuclei type (^{1}H, ^{13}C, etc.); j – is a number of shifting reagents; k – is a number of nuclei of each type; N – is a total number of nuclei; W – is a value, reciprocal to square of the experimental error. The criterion $\boldsymbol{R}^*$ is limited by the inferior limit which is equal to one.

It can happen that some isomeric calculated structures will have rather close values of the $\boldsymbol{R}$ factor. In such a case, a question arises whether the choice of the "best" structure on the basis of minimum $\boldsymbol{R}$ value is justifiable. It has been proposed to use the Hamilton test for this purpose [257]. The ratio $R = R_1/R_2$ is considered, where $R_1 > R_2$. If the value of this ratio exceeds a certain tabular value (at given significance level), this gives grounds to consider structure 2 statistically more significant in comparison with structure 1. However, it has been noted [258] noted that application of the Hamilton criterion in structural studies (the NMR experiments involving the method of lanthanide shifting reagents) is not reasonable since the value of pseudo-contact shift is not a linear function relative to the parameters r, θ. In the same paper, it has been offered to use two known tests (references [35,36] in the work [258]) for the above dilemma, alternative to the Hamilton criterion. Validity of this approach is confirmed [258] by the analysis of experimental paramagnetic shifts in the PMR spectrum of compounds, which structure is shown below.

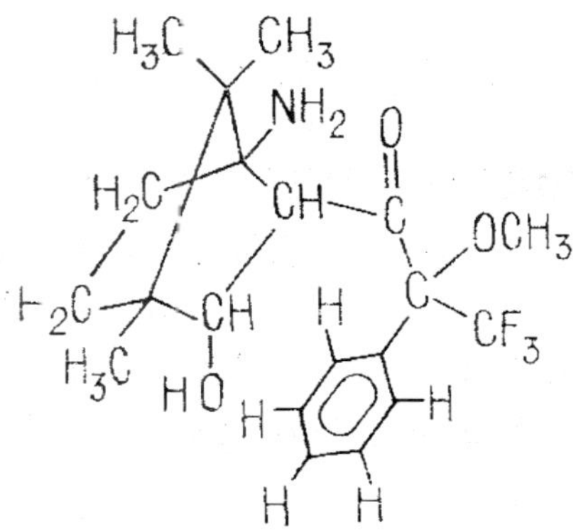

The results of such analysis were known in the literature (before the publication of paper [258]). The problem was in elucidation of predominant conformer for the mentioned compound using LSR. However, the attempts to find clear-cut solution of this problem failed. It turns out that the Hamilton test does not allow distinguishing statistically close spatial isomers. This has been done in the work [258], though the authors conclude that the strict solution of the problem should be based on essentially different model.

It is amply evident that the structural analysis of such molecule represents a problem for any method, including the paramagnetic additives method. This is confirmed by an example of LSR application in structural studies [258]. When investigating the problem concerning preference of spatial structures in conditions of their close statistical importance, it is expedient to analyze the way which has led to such situation. Probably, application of the assumptions based on data of other methods or related to similar (model) compounds in the calculations, will allow one to eventually obtain the adequate information on molecular structure of the studied compounds.

It has already been mentioned that a strict solution of the equations (2.34) is conditioned by the inequality (2.35). Let us consider an example. The VIM molecule has six nonequivalent nuclei (protons), therefore,

$$6m > 2m+24;\ 4m \geq 24;$$

i.e. $m \geq 6$. Thus, to strictly solve the equations (2.33) using paramagnetic shifts of the specified six protons, one should record the PMR spectra with additives of at least six different lanthanides. If to use also the paramagnetic shifts of carbon nucleus signals (five) of this molecule,

$$11m \geq 2m+44; 9m \geq 44, \text{ i.e.}$$

i.e. $m \geq 5$. Prerequisite for this is isostructural character of all employed lanthanides and independence of $\boldsymbol{A}_i$ constant upon the lanthanide type. For axially symmetric complexes, parameter $\boldsymbol{D'}$ is not applicable and the solution of the equations (2.34) requires the following condition:

$$mn \geq m+4n. \tag{2.37}$$

Further simplification can be performed, if the shifts are pseudo-contact ($\boldsymbol{A}_i = \mathbf{0}$) and, besides, the molecule of a ligand is plain ($z_i = \mathbf{0}$), therefore,

$$mn \geq m+2n. \tag{2.38}$$

Thus, using the approach based on minimization procedures and being limited by reasonable assumptions, it is possible to obtain quite reliable structural information, even if the volume of experimental data is lower than it is required by the equation (2.34). Among the reasonable assumptions are the following:

1) Inalterability of the substrate structure in the course of complex formation; this decreases the variables number by 3(n-1), since in this case one should find the position of only coordinating center of the molecule;
2) Lack of the contact contribution to proton signal shifts.

In particular, for VIM molecule (upon condition of such limitations), the unknown parameters will include: three coordinates of coordination nitrogen atom; two parameters defining tensor anisotropy of the magnetic susceptibility of paramagnetic ion $\boldsymbol{D}_j, \boldsymbol{D}'_j$; five constants for $\boldsymbol{A}_i$ carbon nuclei. Therefore, the inequality (2.35) for such a case will be presented as follows

$$(3+2+5)m > 2m+3$$

i.e. $m \geq 1$. Thus, in the conditions of the above limitations, it is sufficient to use additives of one lanthanide to obtain quite reliable structural information.

Scheme LI shows the parameters $\boldsymbol{r}, \boldsymbol{\theta}, \boldsymbol{\Omega}$ of the expression (2.34) and characterizing a relative disposition of the resonating nucleus and the central ion (Ln).

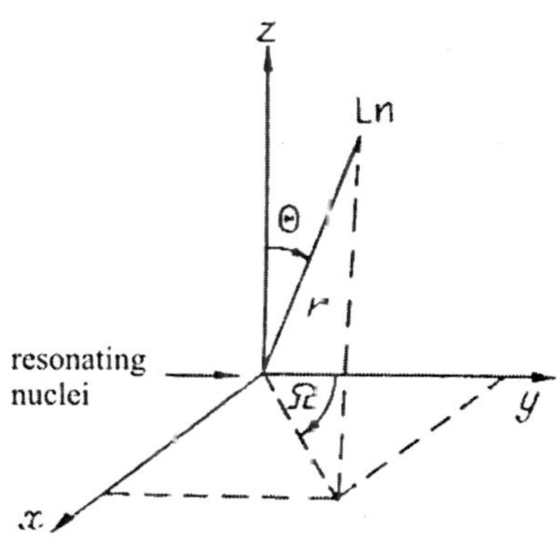

LI

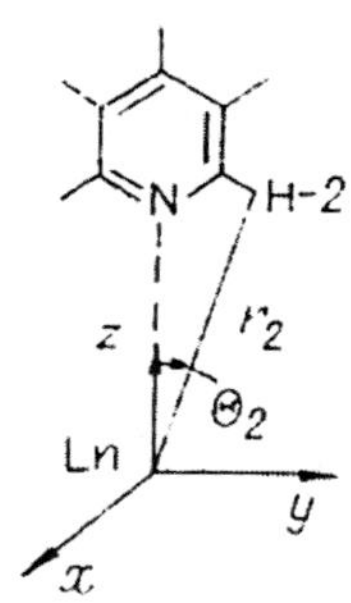

LII

In the system of coordinates shown in Figure LI, the main magnetic axis is directed along the axis z. Usually (though, of course, not necessarily) the origin of coordinates combines with Ln, and the main magnetic axis z is directed along the line connecting Ln to atom, with which the substrate molecule coordinates with paramagnetic ion (Scheme LII). In general case, the axis z (from which angle $\boldsymbol{\theta}$ is calculated) will not necessarily coincide with the direction Ln – coordination center. Similarly, Ln will not always be located in the plane of the ligand molecule (i.e. it is not mandatory that $\boldsymbol{\Omega} = \boldsymbol{0}$). These data are specified in each individual case and, first of all, via determination of the $\boldsymbol{R}(\boldsymbol{R}^*)$ factor. Certainly, such situation complicates the procedure of $\boldsymbol{\omega}_{\exp}$ and $\boldsymbol{\omega}_{their}$ comparison and, hence, the obtaining adequate information on molecular structure. According to opinion of the authors of this monograph, in similar situations it is important to use correctly already available data related to the problem studied (i.e. information on symmetry type of the complex; spectral data obtained by other methods apart from NMR; theoretical data including quantum-chemical calculations, etc.) and on the basis of these data the corresponding conclusions can be drawn. Additional information is especially valuable, when other (except $Ln(DPM)_3$ and $Ln(FOD)_3$) new paramagnetic additives are used. For example, it has been shown that the optical spectroscopy can be successfully applied for the solution of problem related symmetry type of the coordination paramagnetic ion surroundings (see [262] and the references cited therein). Apart from the above-discussed, other techniques, connected with the application of lanthanide shifting reagents [263 - 266], can be effective for the analysis of NMR spectra including when studying the biological systems [263].

Processing of the experimental data in the course of paramagnetic broadenings application in structural studies is similar to the procedure used for shifts, i.e. the experimental values of broadenings (or relaxation times) are compared with theoretically calculated ones. The experimental value of broadening is found by comparison of the spectra recorded with and without paramagnetic additives (the same technique is used for evaluation of paramagnetic shifts). Here it pertinent to note once again that the broadening depends only on r^6 distances to the coordination ion. Therefore, under the conditions of fast exchange and absence of contact contribution, very often structural information can be obtained at qualitative level from the NMR spectra transformed by complex formation, i.e. there is no need to carry out quantitative measurements of substrate and additive mass.

Unlike complexes of rare-earth elements, the predominant character of pseudo-contact HFC in the complexes formed by iron group elements is not obvious. As a rule, establishment of this fact requires additional investigations. Therefore, application of the shifts induced by paramagnetic ions of the first transitional group should involve separation of pseudo-contact and contact shifts. Usually, for this purpose a method developed by Zayev and Molin [267] as well as Horrocks is used [139]:

$$\omega^{Co} = \omega_k^{Co} + \omega_p^{Co} = K\omega^{\mathrm{Ni}} + P\varphi, \tag{2.39}$$

where ω^{Ni} – are shifts in the spectrum of nickel complex; $\varphi = \left(3\cos^2\theta - 1\right)/R^3$ – is a geometrical factor of McConnell-Roberts on formula; K and P – are proportionality constants, which values are found by combined solution of the equation (2.39) for several nuclei. The "best" values of K and P – can be found by varying θ and R (i.e. to achieve the lowest discrepancy between the calculated and experimental shifts). Applicability of the expression (2.39) assumes insignificance of the ω_∂ value induced by nickel paramagnetic ion and similarity of delocalization mechanisms of uncompensated electron spin in Ni^{2+} and Co^{2+} complexes.

The conducted investigations have shown that nickel chelate $Ni(AA)_2$, forming the complexes of 1:2 structure, induces contact shifts, while $Co(AA)_2$ – generates and both contact and pseudo-contact ones (see, for example, [139,268]). However $NiCl_2 \cdot 4L$ and $CoCl_2 \cdot 4L$ complexes (where L – is imidazole, EIM and VIM) induce (with an accuracy of several percent) only contact paramagnetic shifts [35, 36]. It has also been reported [257] on insignificant contact contribution to the paramagnetic shifts of ^{1}H and ^{13}C signals induced by Co^{2+}. It should be noted that the considerable broadening of signals caused by paramagnetic additives of complexes of iron group elements is the main reason hampering their wide application in practice. Against the background of significant progress of NMR spectrometers, the paramagnetic complexes of the first transitional group should find much wider application for the analysis of NMR spectra [269].

Some characteristics of a series of paramagnetic ions of rare-earth elements and iron group are summarized in Tables 2.10 and 2.11. The same tables contain formulas for calculation of pseudo-contact and contact component of the paramagnetic shifts. These formulas are applicable for the cases, when there is information on magnetic properties of the paramagnetic ion as well as on mechanism of contact HFC in the complexes formed by these ions. The value ρ_s – is a spin density of unpaired electron on the corresponding $s-$ orbital, $A-$ constants of contact HFC caused by one electron occupying the corresponding $s-$ orbital. For the most widespread resonating nuclei, these constants are the following (Hz) [178]: 508 (^{1}H), 1119 (^{13}C), 557 (^{14}N), 1659 (^{17}O), 17160 (^{19}F), 3676 (^{31}P). The formula for $\omega_{\hat{e}}$ value calculation is applicable to paramagnetic ion with insignificant spin-spin coupling of electrons. Other designations in the formulas (see notes to Tables 2.10 and 2.11) are common: $k-$ is Boltzmann constant, $T-$ is absolute temperature, β_N – is a nuclear magneton, γ_e and γ_N – are gyromagnetic ratios for electron and N nucleus, $g-$ is an electron g-factor, $\beta-$ is Bore magneton. *g*-Factors for $^{59}Co^{2+}$ and $^{63}Cu^{2+}$ ions are also significantly anisotropic ($g_{\perp}$ and $g_{\parallel}$ considerably differ from each other), therefore, its value for the above ions is not given in Table 2.11. For example, $CuCl_2$ $4VIM$ $g_{\parallel} = 2,33$ and $g_{\perp} = 2,04$.

Relaxation processes in the paramagnetic complexes, which are used as paramagnetic additives, are characterized by the following parameters. Nuclear relaxation times (at least, for ^{1}H and ^{13}C) under the action of paramagnetic ions can decrease to $10^{-1} - 10^{-3}$ s [5, 252 – 254, 270]. Relaxation times of electron spin (τ_s) in the complexes of rare-earth elements are about 10^{-13} s, except for Gd^{3+} (> 10^{-10} c). It has been found [252] that in the Ln(DPM)$_3$·2L (Ln = Pr, Eu, Tb, Dy, Ho, Tm, Yb;L = 3,5-lutidine) complexes, τ_s is constant (~ $7,8 \cdot 10^{-13}$ c), when Ln changes. This time (due to its low value) determines the relaxation times τ_e and τ_c and characterizes the contact and pseudo-contact HFC in the complex, respectively, whereas other times (τ_r and τ_M, see expression 2.6) are significantly higher. For example, τ_r value for the above complexes is about 10^{-10} s [252]. As for paramagnetic ions of the first transitional group, their τ_e and τ_c values is defined in most cases by τ_r time.

Table 2.10. Some characteristics of paramagnetic ions Ln^{3+} [135, 271]

Ion	Z_N	State	S	L	J	g_J of basic state	F	G×F	$\langle S_z \rangle$ at300 K	*a*	$b \cdot 10^4$	$\langle r^2 \rangle$, at.un.	Ionic radius, nm
$^{141}Pr^{3+}$	59	$4f^2$	1	5	4	4/5	–0,0210	–20,7	–2,939	–3,2000	8,7095	1,086	0,100
$^{149}Sm^{3+}$	62	$4f^5$	5/2	5	5/2	2/7	+0,0413	+0,943	0,307	–	–	0,883	0,097
$^{151}Eu^{3+}$	63	$4f^6$	3	3	0	0	–	–	7,658	–	–	–	0,097
$^{157}Gd^{3+}$	64	$4f^7$	7/2	0	7/2	2	0	–	31,500	31,5000	0,0	0,0785	0,094
$^{165}Ho^{3+}$	67	$4f^{10}$	2	6	8	5/4	– 0,0022	–71,2	22,646	22,5000	4,8650	–	0,086
$^{171}Yb^{3+}$	70	$4f^{11}$	1/2	3	7/2	8/7	+0,0318	+39,2	2,589	2,5714	0,58824	0,613	0,081

Note. $\omega_p = -\frac{G\beta^2}{60(kT)^2}\left[\frac{D_z(3\cos^2\theta - 1) + (D_x - D_y)\sin^2\theta\cos 2\Omega}{R^3}\right]$, where $G = g^2 J(J+1)(2J-1)(2J+3)$,

$F = \langle J \| \alpha \| J \rangle$, $D_x = \langle r^2 \rangle F(A_2^2 - A_2^0)$, $D_y = \langle r^2 \rangle F(-A_2^2 - A_2^0)$, $D_z = \langle r^2 \rangle F(2A_2^0)$,

$\omega_k = -A\langle S_z \rangle / g_N\beta_N$, where $\langle S \rangle = a + bT$.

Table 2.11. Some characteristics of paramagnetic ions M^{2+} [135, 271]

Ion	Z_N	State	S	L	g	Ionic radius, nm
$^{55}Mn^{2+}$	25	$3d^5$	5/2	0	2,0014	0,091
$^{59}Co^{2+}$	27	$3d^7$	3/3	3	–	0,078
$^{61}Ni^{2+}$	28	$3d^8$	1	3	2,1728	0,074
$^{63}Cu^{2+}$	29	$3d^9$	1/2	2	–	0,080

Note. $\omega_p = \dfrac{\beta^2 S(S+1)}{27kTR^3}(g_{\parallel} - g_{\perp})(g_{\parallel} + 2g\perp)(3\cos^2\theta - 1)$.

Ligands field: axial symmetry $g_{xx} = g_{yy} = g_{\perp}$, $g_{zz} = g_{\parallel}$;

Octahedral and cubic symmetry $g_{xx} = g_{yy} = g_{zz}$.

$$\omega_k = -A\frac{\gamma_e g_e \beta S(S+1)}{3kT\gamma_N}.$$

It has been reported [151,268] that in the studied complexes, the following values are found: $\tau_{\tilde{n}} = 5\cdot10^{-9}$ s (Cu^{2+}), $5\cdot10^{-11}$ s (Mn^{2+}), $8\cdot10^{-11}$ s (Fe^{3+}), $1{,}6\cdot10^{-12}$ s (Co^{2+}). This example illustrates the common opinion that among elements of the first transition group, Co^{2+} can be employed as a shifting paramagnetic ion. Cu^{2+} and Mn^{2+} have the highest τ_s (~ 10^{-8} s) value, which characterizes relaxation of the electron spin. Mn^{2+} represents an analog of Gd. Co^{2+} has the lowest τ_s value (~ 10^{-12} s) in the series of paramagnetic ions of the iron group elements [158]. As for lifetimes of the substrate molecules in the first coordination sphere of paramagnetic ion (τ_M), it is usually significantly higher than τ_s and τ_r, though in the conditions of fast exchange it can decrease to 10^{-3} s [158, 270] or even to 10^{-5} s [100].

Chapter 3

APPLICATION OF COMPLEX FORMATION PROCESS IN STRUCTURAL STUDIES

In chapter 2 of the present monograph, an efficiency of paramagnetic shifts and broadenings application for obtaining structural information from the NMR spectra transformed by additives of paramagnetic complexes, has been shown. However, in the examples discussed these shifts and broadening play, as a rule, a "passive" role. This chapter (in addition to section 2.4) deals with the application of the NMR spectra transformed by HFC in structural studies combined with theoretical estimations of paramagnetic shifts and broadenings of signals.

3.1. SPATIAL STRUCTURE OF 1-VINYLIMIDAZOLE

According to the current theoretical notions, the presence of lone pair of the pyrrole nitrogen atom in total $\pi-$system of the cycle should stabilize the planar configuration of VIM molecule. Therefore, one can expect a co-planar disposition of the vinyl group and heterocycle, which ensures conjugation of their $\pi-$systems. At the same time, owing to rotation around the $N_1 - C_6$ bond without or with violation of its planar structure, the violation can occur, in particular, for the following reason.

When coplanarity between the imidazole cycle and substituent is violated, N-1 atom, which tends not to break the overlapping of orbitals of its lone pair and $\pi-$system of the double bond, can also be deviated from the cycle plane. If the vinyl group is deviated from the plane by quite high ψ value (as it has been mentioned above, this is an angle characterizing deviation of this group from the heterocycle plane), nitrogen atom will be inverted in the starting position. After rotation of the vinyl group by $\psi > 90°$, N-1 atom again can be deviated from the heterocycle plane tending to ensure the overlapping of the orbitals of the double bond $\pi-$system and its lone pair. This represents an inversion of the pyrrole nitrogen atom. From the conformational point of view, the transitions between planar LIII and nonplanar LIV, LV structures of the heterocycle molecule should be taken into consideration.

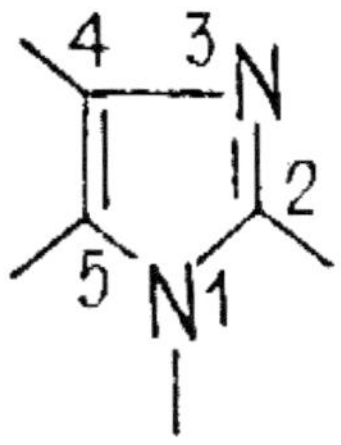

LIII

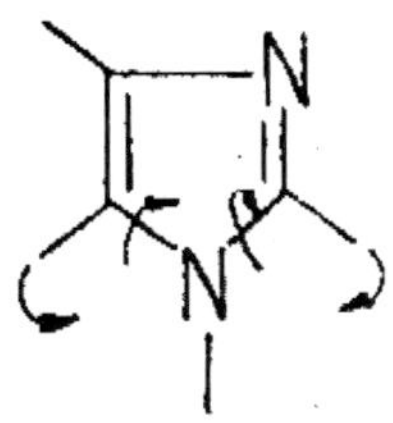

LIV

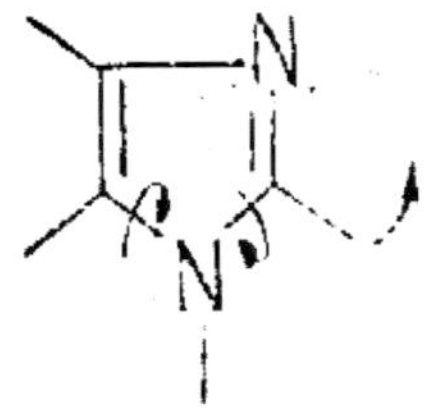

LV

According to the above, two options are possible here.

1. Lone pair of N-1 nitrogen atom is quite effectively conjugated with the ring $\pi-$ system, while conjugation with p_z-electrons of the double bond does not compensate the energy of steric interactions of H-2 and H-5 protons with the vinyl group (if the latter has planar configuration). Therefore, the value of ψ angle should be defined by this interaction, and the pyrrole nitrogen (N-1) will not be inverted.
2. Due to the competition of two $\pi-$systems for conjugation with N-1 lone pair, in the VIM molecule, an inversion of nitrogen atom in the position 1 with deviation of $N_1 - C_2$ and $N_1 - C_2$ bonds as well as deviation of H-2 and H-5 from $C_3 – N_3 - C_4 - C_5$ plane in one (LIV) or different directions (LV) can take place.

To check the experimental data on non-planar structure of the imidazole cycle, the PMR spectra of VIM, recorded with additives of lanthanide shifting reagent $Eu(DPM)_3$, have been studied. The study involves comparison of the induced ω_p and calculated ω_t pseudo-contact paramagnetic shifts [272]. A criterion of comparability of ω_δ and ω_t is the agreement factor R. It is assumed that the ω_t value is defined by McConnell – Robertson formula (see expression 1.25), which can be presented as follows:

$$\omega_t = f(T)\frac{3\cos^2\theta - 1}{r^3} = K\varphi(r,\theta). \tag{3.1}$$

Here $\theta-$ is an angle between the main magnetic axis and the direction to resonating nucleus; $r-$ is a distance from coordination ion to resonating nucleus; K is a constant for the studied sample at temperature T.

To calculate ω_t values, geometrical parameters of unsubstituted imidazole have been employed. It is supposed that the heterocycle of VIM molecule (like unsubstituted imidazole), is planar. The ψ angle varies from 0 to 180°. It has been found that satisfactory agreement between ω_p and ω_t values are reached during the removal of LSR from the coordination center of VIM molecule N-3 on distance r_o which is not less than 4 Å. Here the vinyl group is located in the ring plane. In some papers dealing with the similar calculations, such r_o value is considered as reliable in order to draw conclusions about geometry of both the ligand molecules and paramagnetic complex as a whole. But, according to X-ray diffraction data, the r_0 distance in pyridine complex with Eu(DPM)$_3$ is 2,65 Å [273]. Therefore, it has been noted [63] that r_0 value = 4 Å assumes inadequacy of pseudo-contact model for the description of electron-nuclear interaction in the paramagnetic complex of pyridine with trivalent lanthanides. Besides, it has been shown [258] that at the corresponding procedure of agreement factor R calculation, the distance Eu… pyridine can be diminished to $r_0 \leq 3{,}08$ Å (see also the works [60, 274]). The calculated distance $r_0 \geq 4$ Å in the complex of LSR with VIM is likely overestimated relative to real r_0 value. This specified value is reached when LSR coordinates in the direction lying in the plane of the imidazole cycle and coinciding with α angle bisector ($C_2N_2C_4$), i.e. in the direction of N-3 lone pair orientation. The calculations have revealed that this r_0 value can be decreased. if Eu^{3+} ion is located in the heterocycle plane, but the angle between Eu … N-3 and α angle bisector exceeds 40°.

Such an assumption contradicts the fact that ability of organic compounds to form complexes with LSR is caused by the electric dipolar moment related to lone pair of the ligand heteroatom, which is characterized by certain orientation. Therefore, ω_p and ω_t shifts have been compared [272]. In the study, the following assumptions are made: Eu^{3+} ion is removed from N-3 atom by the distance $r_0 = 2{,}7$ Å along α angle bisector; C-2, N-3, C-5 atoms are located in one plane (in the same where $Eu3^+$ ion is located). It turns out that such conditions allow satisfactory agreement between ω_p and ω_t values to be achieved. For this purpose, it is necessary that (i) N-1 atom is deviated from the cycle plane by 0,4 Å; (ii) the angle $\gamma \approx$ between N_1–C_5 bond and the plane formed by $C_2N_3C_4C_5$ is 15° (Scheme LVI); (iii) H-2 and H-5 protons are deviated from the $C_2N_3C_4C_5$ plane by 5°, but in the opposite directions. The values 0,4 Å, 15° and 5° are obtained without geometry optimization, therefore, they should not be assigned to most energetically favorable conformation of VIM.

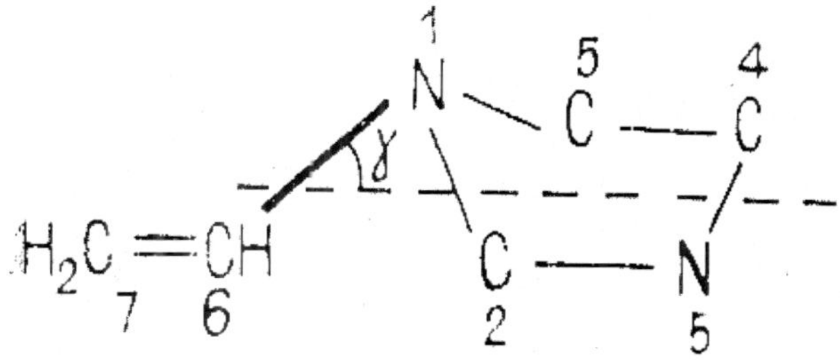

LVI

At the same time these values evidence the reality of non-planar structure of the cycle in the complex. This is confirmed by PMR spectra recorded with LSR additives.

To obtain more detail information on VIM structure, the temperature dependence of ω_p shifts induced by Eu(DPM)$_3$ has been investigated (Figure 3.1) [272]. According to the theory (section 1.3), the paramagnetic shifts induced by this LSR are proportional to T^{-n} where $1 < n < 2$. Notably, for temperatures, at which the experiments have been performed (from – 20º to +60º C), the value n is close to 1. It is suggested [272] that change of temperature can effect both on the heterocycle distortion character and its orientation relative to the vinyl group, and hence on the temperature dependence. The plots shown in Figure 3,1 are divided into two right lines at temperature ~ 5º C. Below this temperature, $n > 2$, above $n = 2$ for ω_2, ω_4 and $n < 2$ for H-5, H_X, H_A and H_B proton signal shifts, i.e. n comes close to the theoretical value.

Additional information on the spatial structure of VIM molecule can be retrieved from the temperature dependence of contact paramagnetic shifts $\delta'_к$ induced by Co^{2+} and Ni^{2+} by in the PMR spectrum of VIM. As a rule, these shifts are proportional to T^{-1}, i.e. they obey to the Curie law. Like in the case with ω_p shifts, it is assumed that alteration of VIM spatial structure can influence the character of uncompensated electronic density delocalization, i.e. the character of dependence $\delta'_к = \delta'_к(T)$. The results of these experiments are depicted in Figure 3.2. The temperature varies within 90 – 170º C (dimethylsulfoxide is used as a solvent). This temperature range is chosen to decrease the line width that allows establishing positions of these lines with sufficient degree of accuracy.

The $\delta'_k = \delta'_k(T)$ dependence for shifts of the ring proton signals has been studied. Due to low absolute values of the vinyl group proton shifts, their temperature dependence is considered. The range of temperature change for the studied sample T differs from that of the LSR. However, in both cases, $\delta'_k = \delta'_k(T)$ function does not depend upon exchange processes and reflects only the nature of paramagnetic ion and ligand that makes it possible to compare the temperature dependence of the shifts induced by diverse paramagnetic ions.

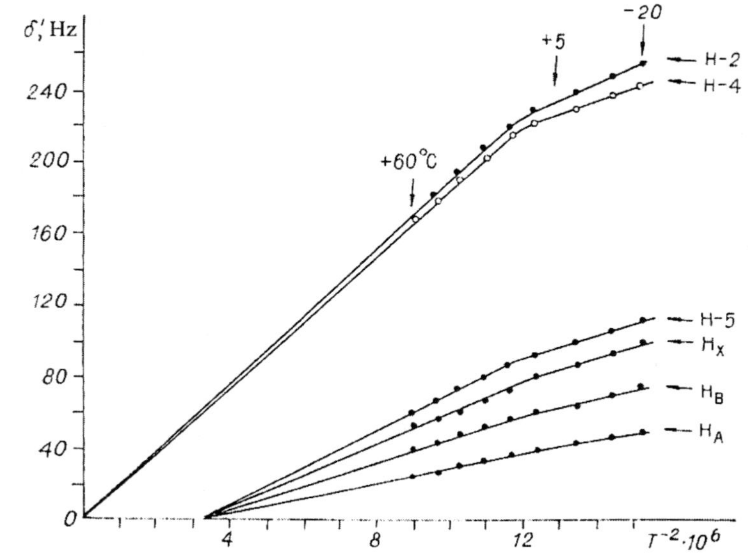

Figure 3.1. Temperature dependence of pseudo-contact shifts induced by $Eu(DPM)_3$ in the PMR spectrum of VIM.

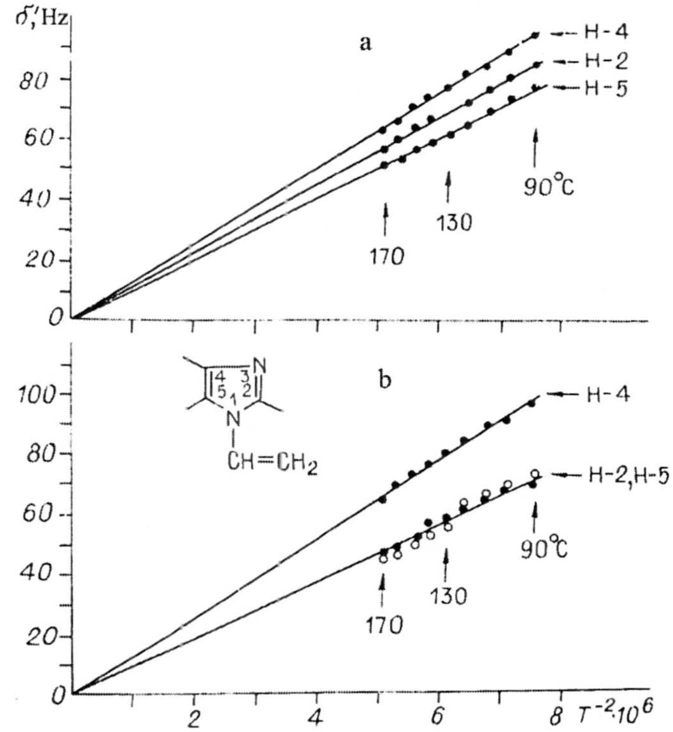

Figure 3.2. Temperature dependence of the contact paramagnetic shifts induced by Ni^{2+} (*a*) and Co^{2+} (*b*) paramagnetic ions in VIM PMR spectrum.

Figure 3.2 shows that δ'_k value for H-2, H-4 and H-5 proton signal shifts is inversely proportional to T^2, i.e. the condition of direct proportional dependence between δ'_k and T^{-1} is not fulfilled. This can be caused by two reasons: (i) approximate applicability of the Curie law for the description of contact hyperfine coupling in the studied complexes; (ii) alteration of spatial and electronic structure of the VIM molecule upon temperature variation. The similar type of $\delta'_k = \delta'_k(T)$ dependence for cobalt and nickel complexes throws discredit upon the fact that violation of the Curie law can be explained only by the first of these reasons. Moreover, temperature dependence of pseudo-contact shifts ω_p testifies to alteration of the spatial structure of VIM molecule in the studied range of temperatures. This results in the change of θ and r (which define ω_p size) values as well as conditions of the unpaired spin density delocalization that leads to appearance of the contact shifts δ'_k. Thus, VIM molecule, coordinated in a solution with either LSR or paramagnetic ions of the first transitional group, becomes significantly non-planar. Eventually, this is reflected in character of contact and pseudo-contact interaction of the VIM molecule nuclei unpaired electrons of the central ion.

The LSR used in the work [272] belongs to the compounds which are characterized by high stereochemical mobility [275]. Besides, it has been reported [266] that the enthalpy of complex formation of VIM with paramagnetic ions approximately equals to 10 kcal/mol.

Therefore, it is not excluded that non-coordinated VIM molecule is also non-planar. The doubling of VIM signals upon decrease of the studied sample temperature could be a direct proof (within the framework of NMR method) of the heterocycle distortion and (or) existence of rotational isomers due to of rotation of the vinyl group around the $N_1 - C_6$ bond. The PMR spectra of VIM have been recorded under these conditions. However, even at – 100° C (solvent – CS_2), doubling of the signals is not observed. Hence, it can be concluded that in the studied temperature range, free rotation of the vinyl group around the $N_1 - C_6$ bond is slowed down. This conclusion is supported by comparison of ω_p and ω_t values, which shows that the H_X proton is oriented towards the H-2 proton. The absence of signals from separate conformers in the PMR spectrum recorded at – 100° C (these signals should exist owing to inversion of the pyrrole nitrogen) testifies that, if even these signals exist, transitions between them at this temperature happen quickly enough (in the NMR scale). Therefore, the spectrum contains the average signals, which position depends on the sample temperature. Variation of temperature in the range of 130° C effects noticeably on positions of the H-2 (0,11 ppm) and H-5 (0,18 ppm) proton signals, and significantly less on positions of the H-4 proton signal (0,02 ppm). Here it is pertinent to note that, according to literature data, nitrogen inversion, for example, in N,N' – dimethylimidazolidine slows down below 140° C.

In connection with a problem of VIM molecular structure, the X-ray diffraction study of $CoCl_2 \cdot 4VIM$ monocrystal in solution is of particular interest [276]. In this work, structure of the imidazole cycle is not specially discussed. It is assumed that this cycle is planar and deviation of C-6 carbons and N-1 nitrogen atoms from its plane does not exceed 0,09 Å, but for the C-5 atom, this value reaches 0,37 Å. However, in $CuCl_2 \cdot 4VIM$, this deviation is not higher than 0,1 Å. X-ray diffraction analysis of $CoCl_2 \cdot$ (1-vinylbenzotriazole)$_2$ complex has revealed that the vinyl groups deviate from the plane of benzotriazole cycles by 13° and 7,5°, respectively [277]. According to the current theoretical notions, stability of conformers is defined by conjugation effects, angular tension, torsion tension, and interaction of indirectly coupled atoms. The combined action of the above reasons also defines, ultimately, the spatial structure of molecules [278]. Certainly, very often one of these factors can play a dominant role. As for aromatic heteroatom compounds (including pyrroles, imidazoles, pyrazoles), the planar structure is rather typical for them [279]. However, its realization, for example, in the cycle of protonated histamine requires compensation of the angular tension as compared with unsubstituted imidazole. This follows from X-ray diffraction data, which show that the geometry of unsubstituted imidazole and tautomeric forms of histamine is different [280]. In $BrFe\{C_{20}N_4H_8(C_6H_5)_4\}$, such compensation does not occur, therefore, the pyrrole cycles, forming porphyrine system, are non-planar [281].

3.2. Complexing Ability and Structure of Si-Substituted Silatranes

Among the achievements of elemento-organic chemistry over the last 30 - 40 years is a discovery of a new class of compounds, silatranes, possessing specific physical and chemical properties [282].

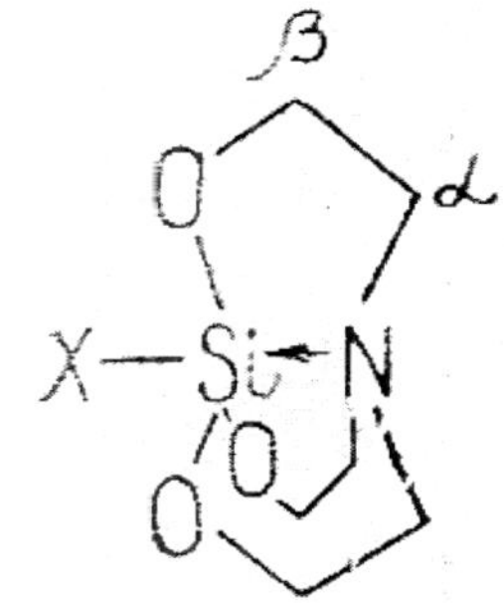

During the first ten – fifteen years after the synthesis of parent representatives of this class of compounds, the silatrane chemistry was progressing intensively (V.A. Pestunovich, V.F. Sidorkin, E.I. Brodskaya, S.N. Tandura, etc.). Nevertheless, in those days it was impossible to explain all the obtained experimental data related to silatranes from general positions. In this regard, the NMR spectra of some Si-substituted silatranes, recorded in the presence of paramagnetic additives, were studied [283 – 285]. It was known that NMR technique is successfully applied for the investigations of HFC in the paramagnetic complexes formed by nitrogen-containing compounds. The confidence that the NMR study of electron-nuclear interaction in the silatranes will allow to obtained new information on this class of compounds, was based on two grounds. First, it is a high sensitivity of pseudo-contact shifts to change of a molecular structure. Second, it is a dependence of contact HFC upon structure of the lone pair orbital, through which uncompensated electronic spin density is localized on the ligand molecule. The results of PMR spectra of some Si-substituted silatranes, transformed by contact and pseudo-contact HFC, are discussed below.

Proton magnetic resonance of Si-substituted silatranes in the presence of Ni^{2+} and Co^{2+}. The PMR spectra of silatranes $XSi(OCH_2CH_2)_3N$, where X = Cl (LVII), CH_2Cl (LVIII), CH_3 (LIX) were analyzed. Spectra of compounds (LVII-LIX) containing an additive of the paramagnetic complex of MCl_2 with EIM, M=Ni (LX) and Co (LXI) were recorded. Addition of $NiCl_2$ or $CoCl_2$ salts to solution of the studied silatranes does not result in the PMR signal shifts, which evidences the absence of complex formation. However, after addition of $NiCl_2{\cdot}4EIM$ or $CoCl_2{\cdot}4EIM$ paramagnetic complexes, the width and position of the PMR signals change, thus testifying to localization of the silatrane molecule in the coordination sphere of paramagnetic ion. This is explained by lability of the added paramagnetic complex in solution:

$$MCl_2{\cdot}4EIM \rightleftarrows MCl_2{\cdot}3EIM + EIM. \quad (3.2)$$

Existence of the equilibrium (3.2) indicates a probability of the following process:

$$MCl_2{\cdot}3EIM + XSi(OCH_2CH_2)_3N \rightleftarrows MCl_2{\cdot}3EIM{\cdot}XSi(OCH_2CH_2)_3N. \quad (3.3)$$

Thus, the silatrane molecule is localized in the coordination sphere of paramagnetic complex. Table 3.1 shows δ' values, normalized relative to signal shifts of α-CH_2-group protons coupled with nitrogen atoms.

Table 3.1. Relative paramagnetic shifts in the PMR spectra of some silatranes and model saturated nitrogen-containing heterocycles

Compound	δ'		Compound	δ'	
	α-CH_2	β-CH_2		$\alpha-CH_2$	β-CH_2
Ni^{2+}ion			*Co^{2+}ion*		
1-Chlorosilatrane	1,00	0,59	1-Chlorosilatrane	1,00	0,29
1-Chlorometylsilatrane	1,00	0,69	1-Chlorometylsilatrane	1,00	0,24
1-Methylsilatrane	1,00	0,63			
Piperidine	1,00	0,47	1-Methylsilatrane	1,00	0,19
Quinuclidine	1,00	0,12			
1-Azaadamantane	1,00	0,77			

The Table contains also the values of δ' induced in the PMR spectrum of piperidine LXII, quinuclidine LXIII and 1-aza-adamantane LXIV by paramagnetic ions Ni^{2+}.

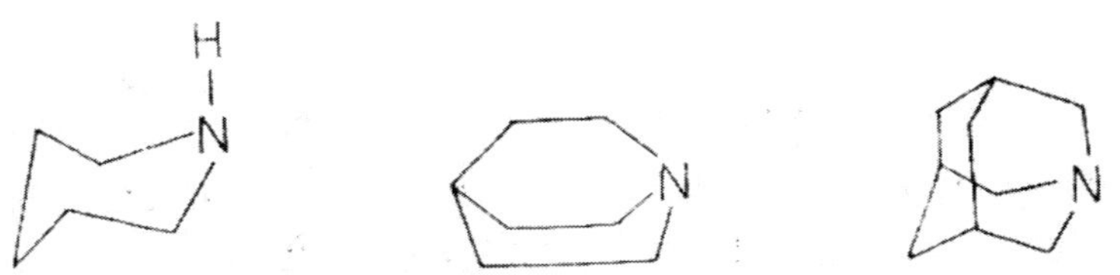

LXII LXIII LXIV

Silatranes show low proneness to coordination, therefore, δ' values are managed to be calculated with sufficient degree of accuracy in the presence of considerable amounts of paramagnetic complex additives (complex/silatrane molar ratio being 1,5 – 2). Usually such δ' values are induced in the PMR spectra of samples containing paramagnetic additive, which amount relative to the ligand does not exceed $10^{-2} – 10^{-3}$. Thus, it can be concluded that only one molecule of silatrane from several hundred is coordinated with the central ion. From Table 3.1 follows that the highest shift (in absolute value) is attributed to signals of the CH_2-group protons. The similar phenomenon is observed for nitrogen-containing heterocycles LXII-LXIV. The $\delta'_\alpha/\delta'_\beta$ ratios in the spectra of LXII, LXIII and LXIV (closer structural analog of silatranes) differ slightly. Thus, the paramagnetic shifts induced in the PM spectra of 1-chlorosilatrane, 1-chloromethylsilatrane and 1-methylsilatrane by Ni^{2+} and Co^{2+}, paramagnetic ions are due to the coordination of these compounds by nitrogen atom.

It is not to be supposed that the only evidence for validity of the above conclusion concerning the coordination center of silatranes is simply the correspondence of paramagnetic shifts in the spectra of these compounds and model nitrogen-containing saturated heterocycles. Such comparison implies the following. By the moment of these investigations, a number of authors have analyzed in detail the HFC mechanism in the compounds, which are used as models. As for the paramagnetic complexes applied as additives, preliminarily, their structure in solution as well as nature of electron-nuclear interaction in them has been investigated [35].

Coordination of 1-methylsilatrane with lanthanide shifting reagents. To study the NMR spectra of silatrane, transformed by pseudo-contact interaction, the paramagnetic shifts δ'_p in the spectrum of 1-methylsilatrane induced by Eu(DPM)$_3$ have been analyzed. For this purpose, δ'_p values are compared with the paramagnetic shifts δ'_t calculated by McConnell – Robertson equation. The comparison of δ'_p and δ'_t values is based on the following assumptions. The molecule of 1-methylsilatrane has four electron-donating heteroatoms, which can be considered as possible coordination centers, and three of them (oxygen atoms) are equivalent; according to molecular geometry of silatranes and LSR coordination to ether oxygen, the europium atom should be located in the plane formed by silicon, oxygen, nitrogen and one of carbon atoms of the silatrane cycle.

The calculation, performed in these conditions, has shown that orientation of the main magnetic axis in the direction determined by Si – O – C bisector angle, leads to satisfactory δ'_t value, only when the distance between Eu^{3+} ion and oxygen atom is not less than 7 Å. This distance for the LSR – substrate complex is unreal, since, according to the X-ray diffraction analysis data, it should not exceed 3 Å. If to take into account that the silicon atom has vacant orbitals corresponding to symmetry of the orbitals occupied by lone oxygen vapors, Eu···O distance can be decreased to 3 Å. Here the main magnetic axis (from which the angle θ is calculated) should pass through the point lying on the middle of distance between Si and O atom. However, the attempts to reach complete agreement between δ'_p and δ'_t values under such condition failed. The further explanation of the observed paramagnetic shifts has been made on the basis of the assumption the 1-methylsilatrane coordinates with Eu(DPM)$_3$ by nitrogen atom. It is assumed that in the calculation of δ'_t values for the case of coordination of 1-methylsilatrane molecule with Eu(DPM)$_3$ by nitrogen atom, the main magnetic axis coincides with the direction silicon – nitrogen. Eu··· N distance varies from 2 to 3 Å.

It has been found that the agreement between δ'_p and δ'_t values is reached, if to accept that 90% of 1-methylsilatrane molecules coordinate by the oxygen atoms, and for other 10% of molecules, nitrogen is the coordination center (Table 3.2). In this case, Eu···O distance is equal to 2,69 Å. The results obtained can be rationalized as follows.

1. Coordination with nitrogen atom is realized in *exo*-conformation of silatrane, which is in equilibrium with *endo*-conformation.
2. Coordination with nitrogen atom occurs without inversion of *endo*-conformation (or with the conformation, which is close to planar).In this case, the association proceeds due to peculiarities of hybridization of nitrogen atom lone pair caused by planarizing of the N(CH$_2$)$_3$ moiety. The calculations performed favor the second hypothesis. Indeed, when the coordination does not involve the transition of *endo*-conformer into *exo*-conformer, the Eu ...N distance equals 2,15 Å. In the case of *exo*-conformation, to reach the agreement between δ'_p and δ'_t values, the Eu... N distance should be 1,5 Å, that is significantly less than usual value. It should be noted that owing to equal

probability of LSR coordination at each of three oxygen atoms, two options are possible.

1. During the localization of 1-methylsilatrane in the coordination sphere of paramagnetic ion, the latter manages to add to each of three atoms (i.e. LSR "is rolled" on them). In this case, calculation of δ_t' value should include averaging of all possible θ and r values, i.e. the expression should be presented as follows:

$$\delta_t' = K \cdot \frac{3\cos^2\langle\theta_i\rangle}{\langle r_i\rangle^3}. \tag{3.4}$$

2. In each coordination, 1-methylsilatrane associates with LSR randomly, but only by one of three oxygen atoms. Therefore,

$$\delta_t' = K \cdot \left\langle \frac{3\cos^2\theta_i}{r_i^3} \right\rangle. \tag{3.5}$$

It has been established that agreement between δ_p' and δ_t' values is reached, if to use the expression (3.5).

Thus, the investigations have shown that only lower part of 1-methylsilatrane molecules coordinates with LSR by nitrogen atom, while with Ni^{2+} and Co^{2+} paramagnetic ions, all the molecules coordinate by this atom. The reality of such situation can be checked by changing concentration of molecules which coordinates with this atom. This can be realized, if to use silatranes having different structure. It is obvious that, according to the expression for pseudo-contact shift, the efficiency of silatranes coordination via the nitrogen atom is defined by the ratio between paramagnetic shifts of proton signals of the protons of N–CH2(α) and O–CH2(β) groups. Table 3.3 shows Eu(ДПМ)$_3$-induced relative paramagnetic shifts of these groups' signals for several Si-substituted silatranes, in which the substituent at silicon atom does not violate a condition of equal probability of LSR coordination across each of three oxygen atoms. Also, this Table contains Si···N distances (known in the literature).

It has been already mentioned that, according to the calculation, only 10% of 1-methylsilatrane molecules coordinate with LSR by nitrogen atom, the ratio $\delta_\beta'/\delta_\alpha' \cong$ being 0,7. This value differs slightly from the similar ratio in the spectra of $HSi(OCH_2CH_2)_3N$and $C_2H_5Si(OCH_2CH_2)_3N$ (see Table 3.3), i.e. molecules of these silatranes coordinate via nitrogen atom approximately with equal probability. However, in $ClSi(OCH_2CH_2)_3N$, the $\delta_\beta'/\delta_\alpha'$ ratio is significantly higher than 1 that testifies to the increase of silatrane molecules, which coordinate with LSR by nitrogen atom, though in this compound, the Si… N distance is the shortest from those shown in Table 3.3. Interestingly, $ClSi(OCH_2CH_2)_3N$ compound is more effectively complexing with $H_3CSi(OCH_2CH_2)_3N$silatrane. Thus, the silatrane with shortest Si ···N distance is the most appropriate compound for coordination via the nitrogen atom. This unexpected experimental fact can be explained, like unusual basicity of

$P(OCH_2CH_2)_3N$, – by repulsive coupling of lone electronic pairs of heteroatom phosphoric fragment [287].Apparently, the decrease of Si ···N distance in the silatranes leads to "pushing out" of lone nitrogen pair, thus increasing its availability to coordination.

Table 3.2.Comparison of relative δ'_p and δ'_t values in the PMR spectrum of 1-methylsilatrane

Values	Nuclei		
	CH_3	OCH_2	$N\text{-}CH_2$
δ'_p	1,00	0,37	0,25
δ'_t	1,00	0,36	0,35
	1,00	0,38	0,26

Note: Above the line – coordination by oxygen, below the line – coordination by oxygen and nitrogen.

Table 3.3.Comparison of δ' values and Si ···N distances in the PMR spectra of some Si-substituted silatranes

Compound	δ'		Si···N, Å
	α-CH_2	β-CH_2	
$HSi(OCH_2CH_2)_3N$	0,79	1,00	
$H_3CSi(OCH_2CH_2)_3N$	0,69	1,00	
$H_5C_2Si(OCH_2CH_2)_3N$	0,64	1,00	2,21
$ClSi(OCH_2CH_2)_3N$	1,32	1,00	2,02

In the connection with problem discussed here, the isomerization type is of special interest. The matter is that transitions between stable isomers occur owing to alteration of the ordinary bond length (see [288], [289] and the references cited therein). This is illustrated by Scheme LXV [288].

LXV

The fundamental moment is that these isomers are characterized significantly by different electronic configuration. Notably, stabilization of one of them is caused by the interaction of orbitals through space, while second isomer is stabilized due to the interaction transmitted through the bonds.

It has been noted above that silatranes can coordinate with $Eu(DMP)_3$ by the nitrogen atom. Such conclusion is drawn from comparison of the calculated and experimental values of the paramagnetic shifts induced by $Eu(DMP)_3$ in the spectrum of 1-methylsilatrane. The agreement between experiment and calculation is reached, when Eu ...N distance is 2,15 Å. Usually, this value is lower in similar calculations. For further discussion, it is necessary to make the following remark. The silatrane molecule contains some heteroatoms with lone pairs, the repulsion between which can quite efficiently influence the spatial structure of the molecule. According to X-ray diffraction analysis data, Si… N distance in the silatrane molecule is noticeably less than sum of Van-der-Waals radii of these atoms. This, in particular, leads to existence of specific coupling between silicon and nitrogen atoms in the silatranes (donor-acceptor type [282]). At the same time, the value of Si···N distance is higher than Si – N bond length.

Summarizing the above, it is possible to assume the following. In the silatrane molecule, a fast alteration of Si ···N distance between two limit values occurs (one for endo-form, second – for covalent bonding). Such alteration should not obligatory leads to that the distribution of electron density will remain unchanged, taking into account the variety of atomic orbitals, which takes place in the silatrane molecule containing some different atoms. Therefore, one can speak about inversion of the electron cloud, which defines structural peculiarities of the silatranes. According to the proposed isomerization mechanism (Scheme LXV), the contribution of each electronic configuration will be defined by the nature of the substituents in cycles and at the silicon atom as well as by intermolecular effects.

Currently, the coordination of silatranes by nitrogen atom as well as Eu···N distance of 2,15 Å, which allows reaching the agreement between δ'_p and δ'_t values, are rationalized as follows. Owing to inversion and rearrangement of electronic structure, such average electronic density can be realized on atoms that electric field of the dipole coupled with nitrogen, ensures localization of the silatrane in the coordination sphere of the central ion.It has already been stated that agreement between the calculated and experimental paramagnetic shifts is observed for the *endo*-form. Transition to *exo*-conformation, i.e. increase of Si···N distance, breaks such an agreement. On the contrary, the decrease of this distance in comparison with distance between silicon and nitrogen atoms in the *endo*-form, i.e. close to Si – N bond length, should not break the agreement. At the same time, Eu···N distance will close to usual values of 2,65 Å.

The results of complex-forming ability of some silatranes discussed in this section of the monograph allow to speak about unusual (from the positions of classical ideas of donor-acceptor interaction) coordination properties inherent in this class of compounds. This inference has been illustrated recently, in particular, by the synthesis of new complex compounds of hypervalent silicon [290].

3.3. Structure of Methyl Thienyl Ketone Oxime Studied by Paramagnetic Additives Method

Oximes of ketones are widely used for the synthesis of diverse pyrroles via the Trofimov reaction [291]. The effect of configurational structure of the initial oximes on the reaction course is one of the fundamental challenges in mechanistic studies. By the moment of publication of the work [292], evaluation of this effect met with difficulties due to the absence of reliable methods for determination of configuration ratio, ambiguity of NMR spectral assignment of E-and Z-isomers showing closely located signals and often transferring one into another. In the work [292], for study of E-Z-isomerism of methyl thienyl ketone oxime LXVI, a starting product in the synthesis of 2-(2thienyl)pyrroles [291], the PMR spectra transformed by shifting reagents have been analyzed.

The spectra of compounds **2,4** have been recorded with $Eu(FOD)_3$ additive. The observed paramagnetic shifts δ' (Hz) are given in brackets.

In the PMR spectra of oxime LXVI, protons of the CH_3-group cause the appearance of two signals (2,33 and 2,28 ppm), ratio of integral intensities being 1:3. The heterocycle protons also give two signals. Addition of $Eu(FOD)_3$, containing a mixture of two isomers of oxime LXVI, leads to the shift of signals of only one isomer (less intensive). Thus, under these conditions, the second isomer does not coordinate with LSR. Therefore, the spectra of two samples have been analyzed. The first sample contains 100% of the isomer having a high-field signal from CH_3-group, and the second one includes 95% of the isomer giving a signal in a lower-field. It has been found that, under these conditions, the high-field isomer also coordinates with LSR that is supported by PMR spectra shown in Figure 3.3. The δ' values are found, when the oxime is used in significant excess relative to LSR. In the PMR spectra of the samples recorded without $Eu(FOD)_3$ additives, signals of H-3 and H-5 protons are overlapped. Therefore, assignment of these signals in the PMR spectra recorded with LSR is a crucial point of the study. This goal is achieved using the spin-spin coupling constants of thiophenes.

To determine the structure of methyl thienyl ketone isomers LXVI, the δ'_t values calculated by McConnell – Robertson formula have been used. Further, these values are compared with experimental δ' values. For calculation of δ'_t values, one should know relative orientation of the substrate and LSR molecules. Usually, such problems are solved on the basis of the statistical analysis of δ'_t and δ' values assuming that structure of substrate-LSR complex is unknown (see section 2.4 of the monograph). At the same time, currently, extensive studies of interaction of the heteroatom molecules containing characteristic functional groups have been executed. In particular, distances from the coordination center of the substrate molecules in LSR are studied in detail. Besides, the spatial parameters characterizing relative disposition of the central ion and ligands coordinated by pyridine nitrogen atom and oxy group oxygen are determined. Moreover, the X-ray diffraction analysis of paramagnetic complexes of LSR with organic compounds has been performed. Therefore, it is justifiable to use the initial parameters, without varying them in the course of comparison of δ'_t and δ' values.

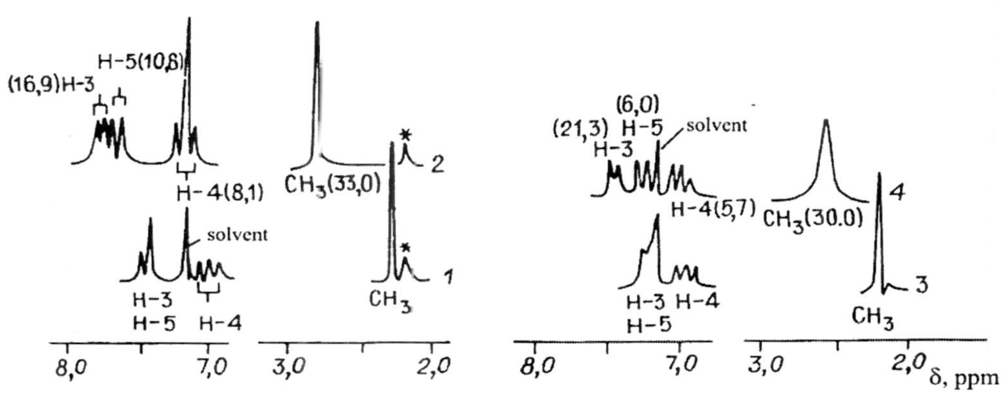

Figure 3.3. PMR spectra of methyl thienyl ketoxime 1, 2 – Z-isomer; 3, 4 – E-isomer.

It has been supposed [292] that the distance r_0 from Eu(FOD)$_3$ to oxime LXVI oxygen and nitrogen should not exceed 2,5 Å and 3,0 Å, respectively. For calculation of δ'_t values, geometry of thiophene and dimethylglyoxime is used. The δ'_t values are calculated for four planar isomers assuming that isomers of oxime LXVI are coordinated with Eu(FOD)$_3$ either by nitrogen or oxygen atom. From all the calculated structures, satisfactory agreement between δ'_t and δ' values is obtained only for two (E and Z) isomers. In this case, the experimental paramagnetic shifts, observed in the spectrum of low-field isomer, coincide with the calculated ones of the Z-isomer provided that it coordinates via nitrogen atom (=3,0 Å, the europium atom is located in the plane of oxime, angle φ = 120°).

E_a E_b

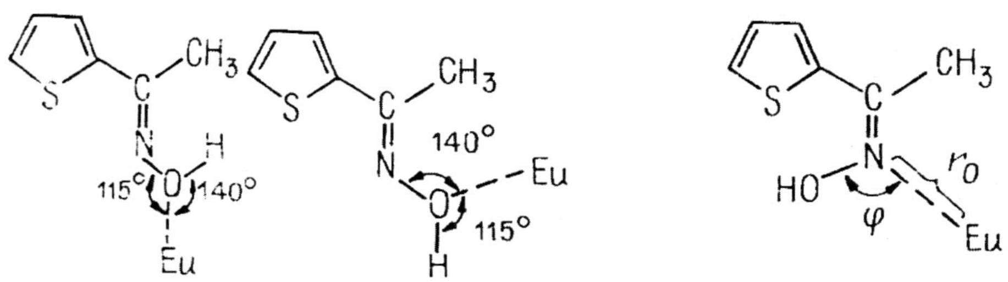

E-isomer

δ'_t :CH_3 – 1,00;H_3 – 0,71;H_4 – 0,23;H_5 – 0,19

δ' :CH_3 – 1,00;H_3 – 0,71;H_4 – 0,29;H_5 – 0,20

Z-isomer

δ'_t :CH_3 – 1,00;H_3 – 0,50;H_4 – 0,29;H_5 – 0,33

δ' :CH_3 – 1,00;H_3 – 0,51;H_4 – 0,25;H_5 – 0,33

The agreement between δ' values, observed in the spectrum of high-field isomer, and δ'_t values is attained for E-isomer coordinated by oxygen atom provided that from all molecules of the ligand interacting with Eu(FOD)$_3$, 70% are isomer of E_a and 30% are isomers of E_6 (r_0 = 2,3Å). Therefore, δ' values are average.

In this line, the following circumstance is of certain interest. It is found that Eu(FOD)$_3$ significantly broadens a signal of the CH_3-group of high-field isomer (see Figure 3.3). Taking into account the nature of the broadenings caused by paramagnetic additives, this experimental finding can be explained by the fact that in the complex of Eu(FOD)$_3$with E_6 isomer, the methyl group is located noticeably closer to the central ion in comparison with its location in the complex formed by the E_a and Z-isomers.

Thus, oxime LXVI has geometry, characteristic either for Z- or for E-isomer. Noteworthy, the latter is responsible for generation of the corresponding high-field signals the PMR spectra. It should be emphasized that E-isomer of oxime LXVI turns out to be more thermodynamically stable and isomerization proceed at quite low temperatures. For example, at 50° C, the Z-isomer dissolved in CCl_4 is transferred to E-isomer by approximately 50% for

1,5 h. At room temperature, thermodynamic equilibrium corresponds to the ratio of Z- and E-isomers of ~ 1:3.It is also found that Z – E-isomerization is slowed down in $CHCl_3$ (in comparison with CCl_4). Such effect of the medium can be explained by the fact that chloroform, a more polar solvent than CCl_4, stabilizes better more polar Z-isomer.

Later, the results on Z – E-isomerization, obtained here with a help of the paramagnetic additives method, have been confirmed by ^{13}C NMR studies and ab initio quantum-chemical calculations of spin-spin coupling constants for a series of aliphatic, alicyclic, heteroatom and carbonyl-containing oximes as well as azomethines [293-294].

3.4. Peculiarities of Molecular Structure of Nickel Paramagnetic Complexes with Nitroxyl Radicals

As it was already mentioned in section 2.1 of this monograph, the quantitative evaluation of signal broadenings caused by hyperfine coupling is reduced to the solution of the Bloch equations modified by McConnel who included a chemical exchange in them. The analysis of these equations leads to the expression (2.2) for the observed paramagnetic broadening of signals:

$$\Delta = \frac{P_M}{\tau_M}\left[\frac{\left(\frac{1}{T_{2M}}\right)^2 + \frac{1}{T_{2M}\cdot\tau_M} + \omega^2}{\left(\frac{1}{T_{2M}} + \frac{1}{\tau_M}\right)^2 + \omega^2}\right],$$

where τ_M is the duration of resonating nucleus location in the paramagnetic environment; P_M is molar fraction of paramagnetic complex in the sample; ω is a paramagnetic shift of the resonance line; T_{2M} is a time of transverse relaxation of nucleus in the complex. From this equation follows that the Δvalue is the composed function of τ_M, T_{2M} and ω. Depending on nature of the studied complexes and experimental conditions, the scattering of the Δ values can be quite broad that significantly complicates application of the above expression for the analysis of experimental spectra. Therefore, the simplified formulas are usually used for the analysis of the observed Δvalues. Variant of the simplification in each specific case is conditioned by type of the object defining the ratio between $(\tau_M)^{-1}$, $(T_{2M})^{-1}$ and ω, and by character of the solved task. If for example, obtaining structural information is the main task of the study, the most convenient is approximate expression for Δ answering to the condition of fast exchange, for which Δvalue is defined by the expression (2.18) and (2.19).

Vigorous growth of new paramagnetic systems including heterospin significantly expands a circle of urgent structural problems. As a consequence, the application of total expression (2.2), instead of its approximations, becomes crucial for such purpose. It is especially important to use the expression (2.2) in the analysis of NMR spectra of heterospin complexes of paramagnetic ions of transition metals with organic radicals, since for such objects the situation is additionally complicated due to existence of spin-spin coupling between unpaired electrons of the molecule paramagnetic centers. In this case, application of approximate analytical expressions can lead to essential distortion of functional dependencies between parameters.

For better understanding of the experimental ^{1}H NMR spectra related to the Ni(II) complexes with nitroxyl radicals presented below, it is necessary to discuss the results of modeling of the paramagnetic broadenings alteration. The step-by-step analysis of the expression (2.2), including determination of certain values of any parameters or limits of their change, has been performed [295].Figure 3.4 shows functional dependence

$$\Delta = \Delta\left(T,\omega\right), \tag{3.6}$$

where $\boldsymbol{T}$ is a temperature of the studied sample. The three-dimensional diagram was received provided that $P_M = 1$, and τ_M is defined by the common formula:

$$\tau_M = \left(h/kT\right)exp\left[\left(\Delta H - \Delta S\right)/RT\right]. \tag{3.7}$$

The computer modeling is based on the assumption that $\Delta H = 15 \cdot 10^3$ kal/mol, $\Delta S = 20$ kal/mol·K,$T_{2M} = 10^{-2}$ s. These specific values of parameters chosen for modeling are not essential. Nevertheless, they are selected from the range of values, which are quite characteristic [296].Functional dependence of paramagnetic shift on temperature is defined by the expression (1.3).

The graphical representation of the dependence $\Delta = \Delta(T,\omega)$ (figure 3.4) allows finding two important tendencies of the general character, which are typical for the functional dependence discussed. First, change of the temperature leads to a shift of broadening maximum of the resonance line. Second, variation of the temperature results in the increase of both maximum values and range of paramagnetic broadenings of the resonance signal owing to square dependence of Δ value upon paramagnetic shift at approaching to maximum of the paramagnetic broadening.

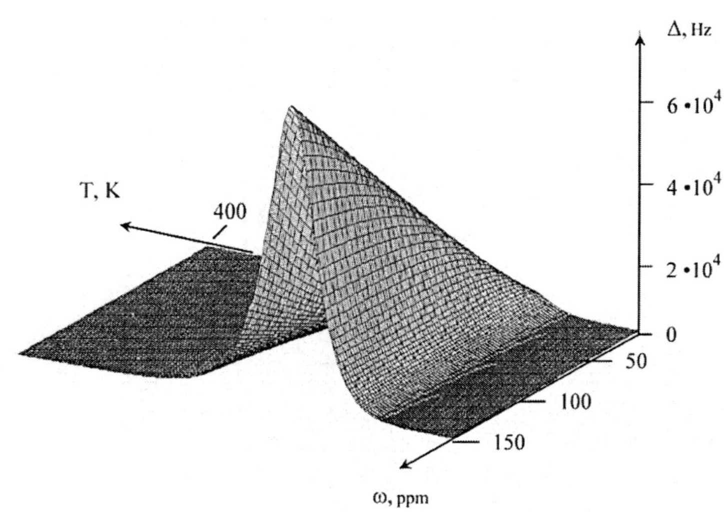

Figure 3.4. Modeling of Δ(T,ω) dependency (ΔH and ΔS are constant).

Further, it is important to pay attention to the following circumstance. Since $d\tau_M/dT <$ 0, at certain temperature (T_0) corresponding to maximum of the paramagnetic broadening, τ_M and Δ values begin to decrease as temperature grows. And, if τ_M increases at a temperature $T_1>T_0$, $\Delta=\Delta(T,\omega)$ function will have another maximum at $T_2 > T_1 > T_0$.From the expression (14) follows that the τ_M value can be increased, for example, via variation of the thermodynamic parameters. Figure 3.5 shows $\Delta=\Delta(T,\omega)$ dependence. Here it is assumed that ΔS value decreases to 10 kal/mol at T_1=300 K. For a specific paramagnetic system, the reason of such decrease should be physically meaningful that requires, of course, separate discussion. Here it is important that the second maximum does appear on the diagram of functional dependence at τ_M variation.

It is obvious that, if at a temperature $T_3 > T_2$ the τ_M value changes again, the diagram of $\Delta = \Delta(T, \omega)$ will show the third maximum, etc. From here it follows rather nontrivial result, namely under the corresponding conditions, the resonance signal, the position of which in the NMR spectra is defined by paramagnetic shift ω, may, in principle, undergo the periodic narrowing and broadening. In other words, the $\Delta = \Delta(T,\omega)$function becomes modulated owing to variation of the τ_M parameter characterizing exchange processes in the system, which HFC is studied. Experimental manifestation of such effect depends both on the HFC parameters and possibility of temperature variation in quite wide range. From physical point of view, alteration of the τ_M value owing to decrease of the ΔS value, observed during the modeling, represents a temporal ordering in the system of nuclear spin (temporal polarization). Actually, the system includes one more channel for energy exchange between resonating nuclear spins and the lattice, including with a subsystem of electronic spins. This, at least, in principle opens new opportunities for manifestation and detection of magnetic resonance effects as well as for influencing the subsystem of nuclear and electronic spins in the presence of HFC.

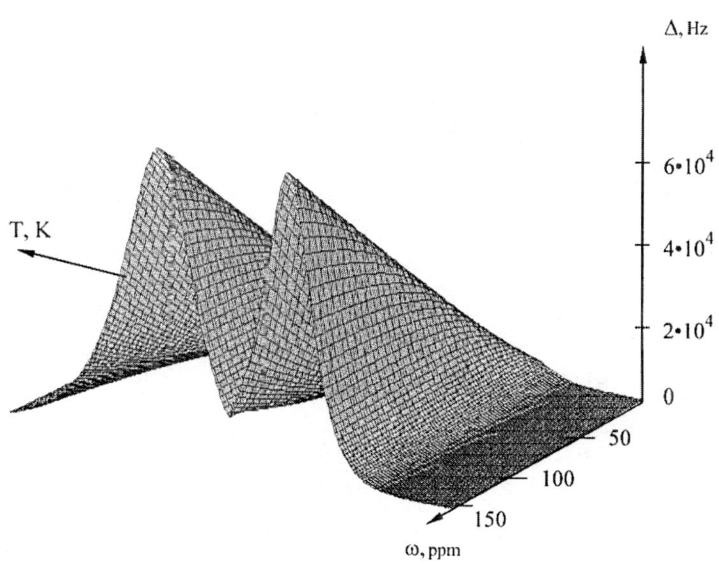

Figure 3.5. The shape of dependence $\Delta(T,\omega)$ at single step-wise change of ΔS.

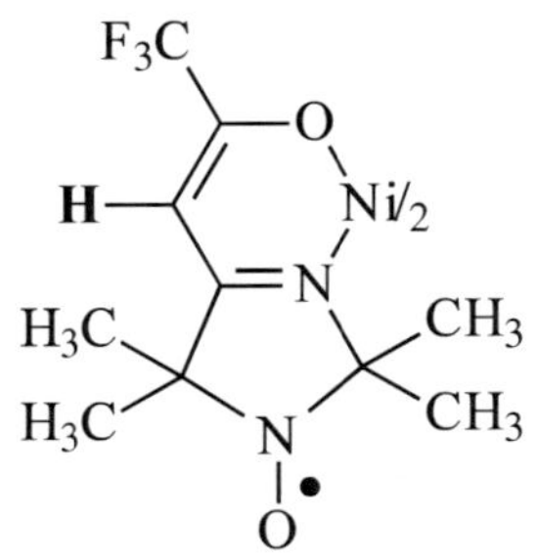

1

Figure 3.6 shows typical transformation of the spectrum of such complex upon varying temperature of the studied sample directly in the sensor of NMR spectrometer.

It should be stressed here that the initial expression (2.2), used for the modeling, has been obtained for the description of so-called chemical exchange, when the resonating nucleus is located in the first coordination sphere of the paramagnetic ion during time τ_M. In the general case, this exchange should be considered as the effect of common character, when the resonating nucleus simply changes its position between the states with different magnetic surroundings. Hence, it follows that from the NMR positions, the following situations are possible:

1. Really, there is an exchange, when the ligand molecule is periodically located in the first coordination sphere of metal paramagnetic ion. As a consequence, this molecule exists in different magnetic environment in different periods of time. In such a case, the observed paramagnetic should represent a linear function of molar concentration of the complex and the dependence $\omega = \omega(T)$should obey to the Curie law.
2. The dynamics described in the item 1 is absent for the ligand. However, the equilibrium between different conformations of the complex (for example, classical square ↔ tetrahedron) takes place. This equilibrium is characterized by various magnetic properties. The experimental dependence $\omega = \omega(T)$ for such process (and similar ones) should be described by the function containing an exponent.
3. The variation of the sample (which NMR spectrum is detected) temperature effects on the spatial structure of ligand molecules of the paramagnetic complex and, eventually, on the character of HFC. It is easy to conclude that in the third case, nature of temperature dependence $\omega = \omega(T)$ is not a priory obvious. For the purpose of experimental check of the modeling performed, the temperature dependence of paramagnetic shift and signal broadening of the =CH-proton of enaminoketone group in the ^{1}H NMR spectrum of bischelate **1** has been investigated:

The experimental dependence $\omega = \omega(T)$ of the =CN–proton signal has complicated character and in the region of high fields does not obey to the Curie law (curve 2 in Figure 3.7). Below 240 K, intensity of this signal decreases, and a new signal appears in low-field. Paramagnetic shift of the latter signal increases upon decreasing of the sample temperature that corresponds to the Curie law (dependence 1 in Figure 3.7).

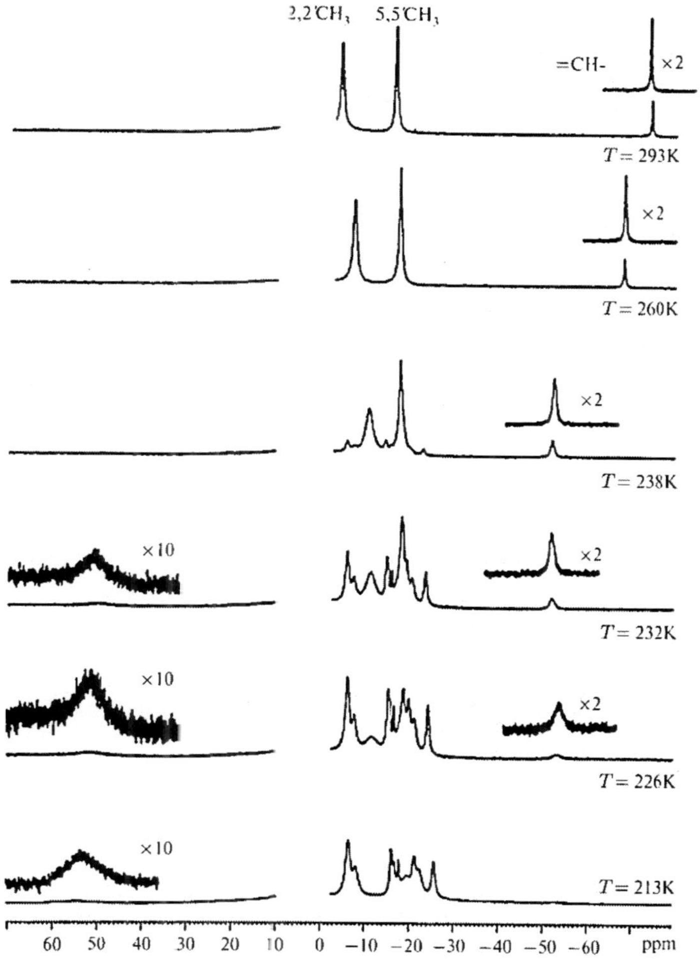

Figure 3.6. Transformation of ^{1}H NMR spectrum of compound **1** in CH_2Cl_2 upon changing the sample temperature.

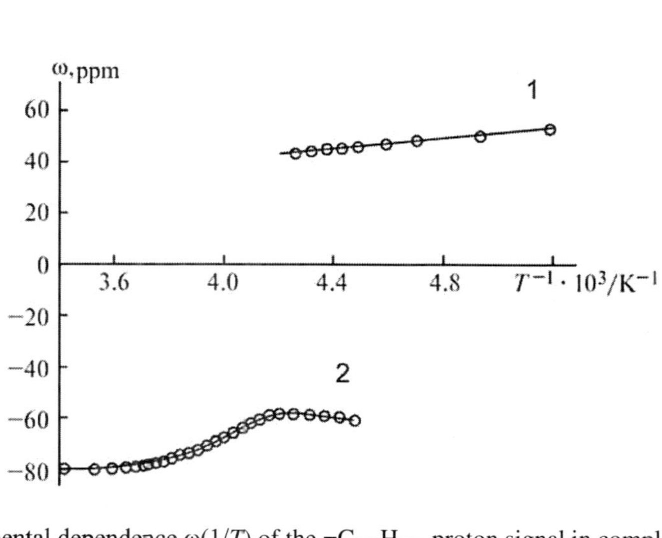

Figure3.7. Experimental dependence $\omega(1/T)$ of the =C—H—proton signal in complex **1**: in low-field (*1*) and in high-field (*2*).

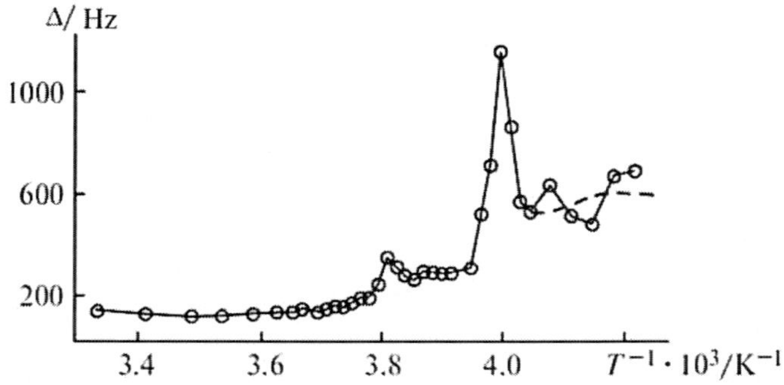

Figure 3.8. Experimental dependence Δ(1/*T*) of proton signals of the =CH — fragment in complex **1.**

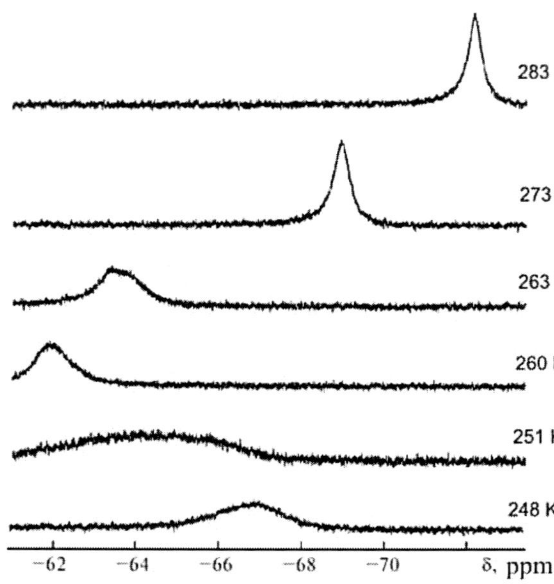

Figure 3.9. Change of proton signal of the =CH-fragment in 1 in the region of high-fields at temperature alteration.

Experimental dependence $\Delta(1/T)$ for the =CH– proton signal also has a complicated character (Figure 3.8). It contains two resolved maxima, thus confirming fundamental correctness of the above-discussed modeling. Actually, Figure 3.8 represents an experimental confirmation of the fact that in similar systems, the NMR signal can undergo the repeated narrowing and broadening upon changing the temperature.

It is pertinent to pay attention to the circumstance that intensity of maximum at 264 K is approximately 3 times lower than that of maximum at 252 K (Figure 3.8). At the same time, on similar theoretical dependence both maxima are identical in intensity (Figure 3.5). This distinction is explained by that fact the diagram in Figure 3.5 is simulated assuming simultaneous (jump-like) alteration of entropy values for all the molecules. In reality, the process of states over-population is expanded on a temperature scale and, therefore, finite time (Δt) is required for the transition from one magnetic environment to another in the course of gradual change of temperature. Generally, each of such states can be characterized by the corresponding set of parameters P_1, τ_1, ω_1, Δ_1 и P_2, τ_2, ω_2, Δ_2. In this case, when ω' ($\omega' = \omega_1 - \omega_2$) value, expressed in frequency units, remains comparable with $1/\pi\Delta t$ value at given temperature, the observed spectrum will be partially average or even contain signals from two states. This explains shape of the = CN-proton signal which is observed in the course of temperature alteration (Figure 3.9). It is shown in dependence diagrams $\omega(1/T)$ and $\Delta(1/T)$ (Figure 3.7 and 3.8, respectively). The reasons of ^{1}H NMR signal transformation in complex **1** have been rationalized [295] using the known data on stereochemical flexibility of this compound [297] allowing to admit its existence in solution in the form of different conformations.

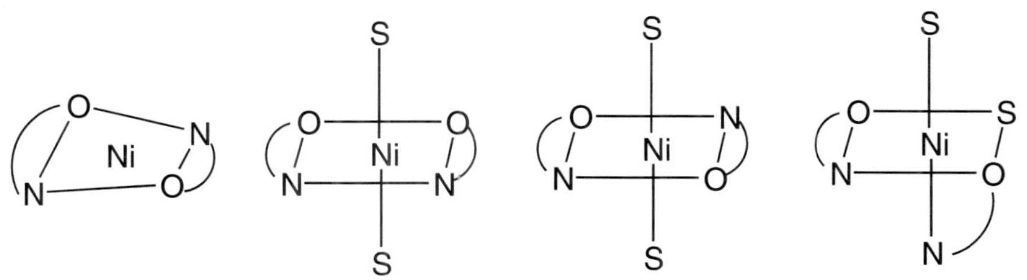

tetrahedron tetrahedron octahedron octahedron

1a 1b 1c 1d

At 283K - 248 K, two signals of the C-H protons can be detected in the spectrum (see Figure 3.9).Noteworthy, the temperature range of these spectra varies only within 1-2 K. A possibility of the NMR detection of such fine effects of heterospin systems is important not only for demonstration of exclusive sensitivity of the method, but for potential control of the molecular states in solution during design of molecular magnetics.

Stereochemical flexibility of complex **1** can be completely suppressed by introduction of the additional bidentate ligand. Actually, the introduction of such ligand into molecule **1** transforms all possible conformations to the only **1d**. In reality, model complex **2** having such configuration has been synthesized by the reaction of bischelate **1** with phenanthroline.

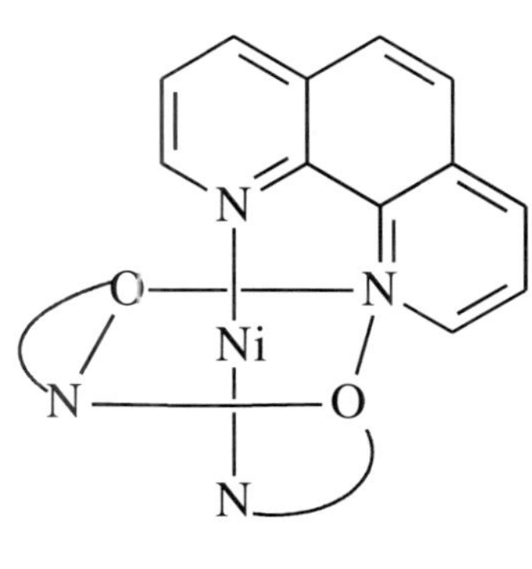

2

Figure 3.10 shows the ^1H NMR spectra of complex **2** recorded at two different temperatures. It is important to note that the signal of the =CH- proton in all temperature range of the measurements is observed in the region of low-fields. This unambiguously indicates that at a temperature below 240 K, the complex **1** in solution accepts conformation **1d**. Figure 3.11 and 3.12 depict the plots of dependencies $\omega = \omega(T)$ and $\Delta = \Delta(T, \omega)$ for the = CN- proton signals in NMR spectrum of complex **2**. Their linear dependence clearly evidences realization of the Curie law and stability (unlike bischelate **1**) of complex **2** upon temperature variation.

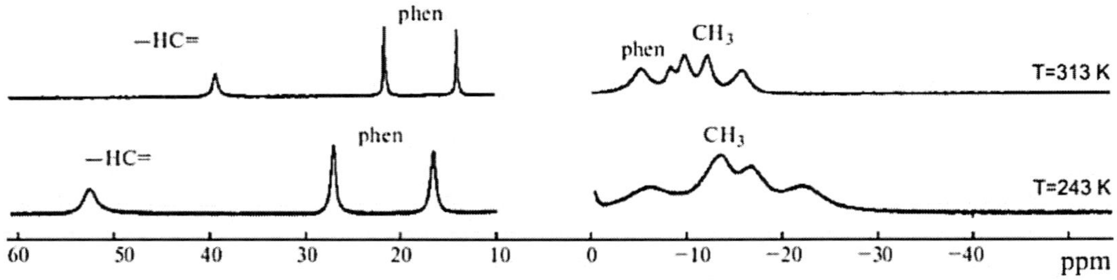

Figure 3.10. ^1H NMR spectra of complex **2** in CH_2Cl_2.

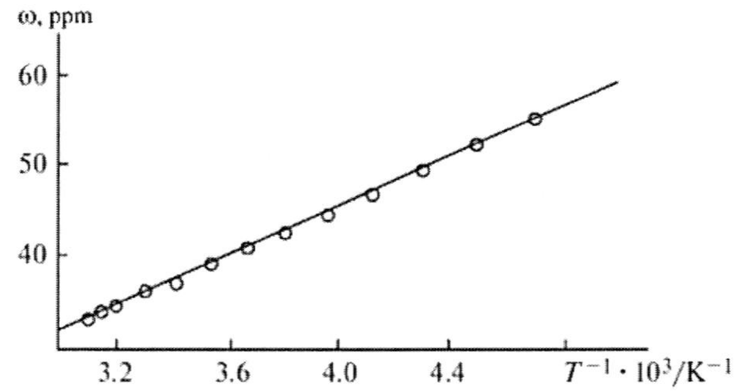

Figure 3.11. Dependencies ω (1/*T*) for proton signals of the =CH— fragment in NMR spectrum of comlex 2.

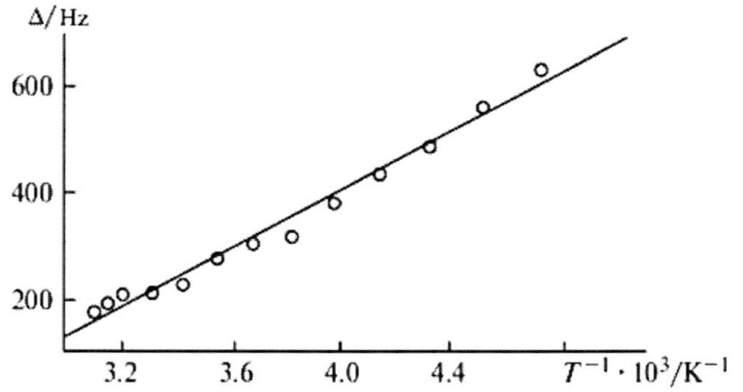

Figure. 3.12. Dependencies ω (1/*T*) for proton signals of the =CH— fragment in NMR spectrum of complex 2.

It is necessary to pay attention to another peculiarity of the NMR spectra of the Ni(II) complexes with nitroxyl radicals. The increase in temperature leads to augmentation (instead of decrease) of the absolute value of paramagnetic shift of the = CN- high-field signal (Figure 3.6) as it is demanded by the Curie law. This effect has been easily explained. Since different paramagnetic centers are present in the complexes with radicals, it is possible to assume that they can induce paramagnetic shifts having different signs. This assumption has been confirmed in the study of temperature dependence of ^{1}H NMR spectrum of specially synthesized complex **3** containing diamagnetic analog of spin-labeled enaminoketone in complexes **1** and **2**. Figure 3.13 illustrates behavior of the =CH- signal upon temperature variation. Like in the case of complex **1**, at high temperatures, the signal of this proton is observed in a high-field.

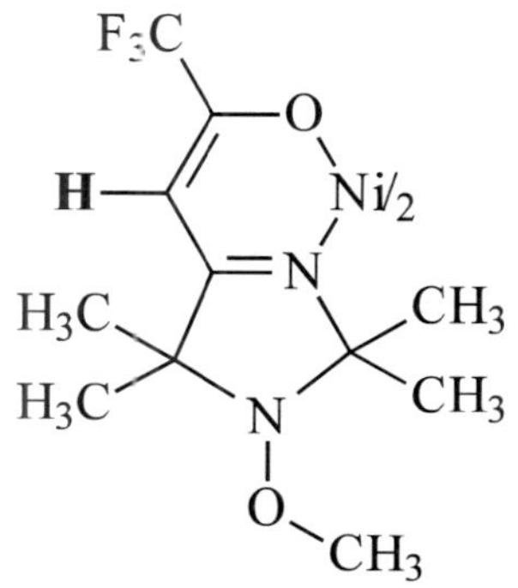

3

At lower temperature, this high-field singlet shows rather complicated behavior due to stereochemical flexibility of molecule **3**. Below 294 K, it is split into two signals, while below 233 K, a higher-field signal first broadens and then (below 203 K) is split into two signals. Thus, the temperature dependence of paramagnetic shifts of all three signals obeys the Curie law, i.e. at temperature decrease, the absolute value of the shift increases (Figure 3.13, 3.14a).

Like for complex with radical **1**, the paramagnetic shifts of signals undergo numerous broadening and narrowing. Therefore, the deviation of dependence $\boldsymbol{\omega} = \boldsymbol{\omega}(\mathbf{T})$ from the Curie law for **1** is indeed caused by induction of paramagnetic shifts with different signs from nickel ions and nitroxyl fragments that defines rather complicatedcharacter of alteration of paramagnetic shift of the=CH–proton in the spectrum of compound **1** upon temperature changing.

Also, it should be underlined that in the spectrum of compound **3**, upon variation of temperature, the signal in low-field is not detected, unlike the spectrum of complex **1** (Figure 3.6). This allows another effect of the paramagnetic ligand in the complex structure on the NMR spectra to be revealed. In heterospin molecule, structural dynamics of stereochemically flexible molecule of the complex upon variation of temperature changes the character of spin coupling. This can influence not only the value of spin density but even its sign (in certain conformations) that is actually observed in the spectrum of compound **1** at a temperature below 240 K. Therefore, the NMR technique makes it possible to determine the temperature regions, where exchange interaction between the paramagnetic centers can have significantly different character. And, if structure of the molecules remains intact upon crystallization from solution, in the solid phase, the exchange interactions will also be significantly different.

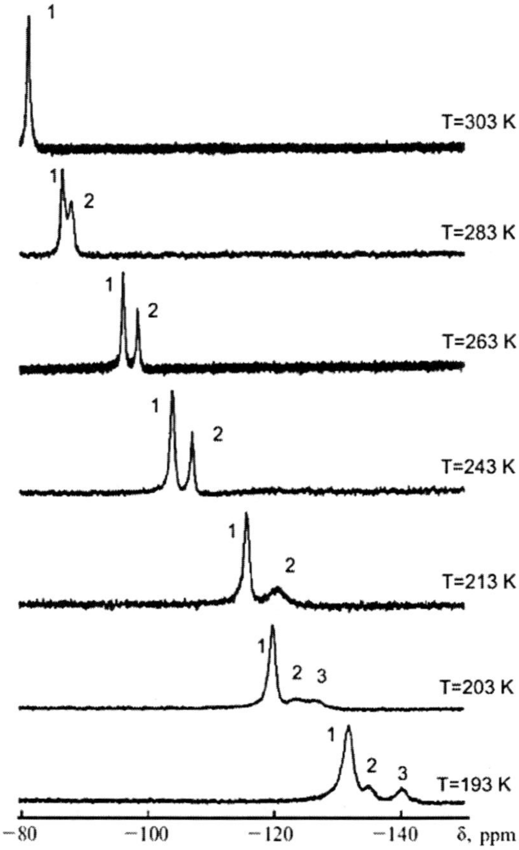

Figure 3.13. Alteration of proton signal of the =CH — fragment of molecule **3** in[1]H NMR spectrum in CH_2Cl_2 upon temperature change.

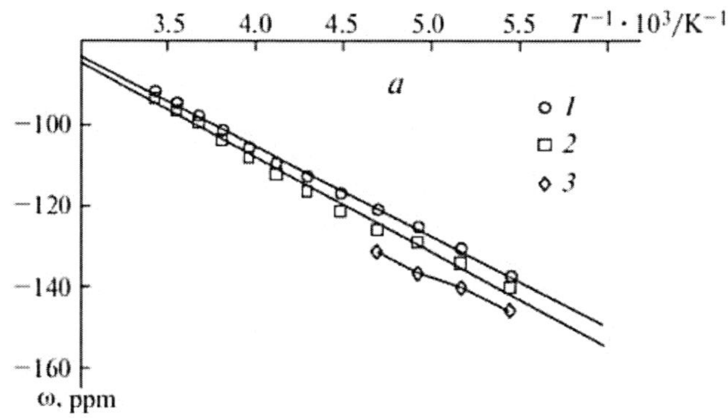

Figure 3.14. Experimental dependences ω(1/*T*) of proton signals of the =CH— fragments for different conformations of molecule **3.**

Thus, the study of hyperfine coupling, modulated by exchange, allows obtaining valuable information on molecular dynamics in solutions of paramagnetic ions of transition metals with stable nitroxyl radicals. To illustrate the directed synthesis of solid phase of the heterospin compound containing structural fragment of certain conformation, the following example will be discussed. It has been reported [297] on the synthesis of solid-phase compound **1** in the form of two polymorphic modifications. Both of them have the layer-polymer structure due to the coordination of Ni^{2+} ion with O atoms of the nitroxyl groups of two neighboring bischelate molecules. However, in one modification (α), oxygen atoms are coordinated in the *trans*-position of coordination octahedron, whereas in another one (β), the coordination occurs in the *cis*-position. Under usual conditions, α-modification is thermodynamically more stable, but β-modification is gradually or quickly (upon heating) transferred into α-modification. Therefore, compound **1** can be readily synthesized in the form of α-modification. The β-modification can be isolated, for example, from CH_2Cl_2 solution only by fast evaporation of the solvent that is accompanied by cooling of the reaction mixture. Upon growing the β-modification monocrystals, the latter are always formed as impurity crystals to the α-modifications, which are separated mechanically. The study of temperature dependence of the ^{1}H NMR spectrum of compound **1** in CH_2Cl_2 has shown that molecules of compound **1** tend to adopt *cis*-octahedral conformation (1d type) at a temperature below 238 K. Indeed, the crystallization of compound **1** from a solution of dry CH_2Cl_2 at a temperature below – 40 °C allow monocrystals of β-modifications to be readily obtained.

3.5. Peculiarities of Intramolecular Exchange and Valence Tautomerism in Metal Semi-Quinolates

Over the last decade, the great research interest has been expressed in the hete-rospin systems synthesized from coordination compounds of paramagnetic ions of transition metals with stable nitroxyl radicals. First of all, this is due to a possibility to obtain detail information on the nature of intramolecular interactions in the multi- electron systems, the specific of which is defined by hyperfine coupling. Ultimately, such information makes it possible to solve fundamental tasks related to the problem of structure and dynamics of molecules. In its turn, the solution of the specified tasks will allow the materials with practically valuable properties to be synthesized.

Among the fundamental problems, intramolecular rearrangements, involving delocalization of unpaired electrons from ligand molecules to the metal (for example, cobalt), thus changing valency of the metal, hold a central position. Such rearrangements are commonly referred to as valence tautomerism [298]. As applied to compounds of the first transition group, these intramolecular processes lead to populationn of 3*d*-orbitals of the metal by unpaired electrons. As a consequence, such compound becomes, in particular, paramagnetic. Delocalization of unpaired electron spins on 3*d*-orbitals results in characteristic shifts and broadenings of signals in the NMR spectra (chapter 1 of this monograph). Therefore, it is possible to study the valence tautomerism on the basis of high resolution

NMR spectra transformed by hyperfine coupling between unpaired electron spins and resonating nuclei.

All the above-discussed has stimulated the detail analysis of ^{1}H NMR spectra of semiquinolate complex $Co(SQ)_3$ (**4**), which is capable of showing the valence tautomerism effect [299].

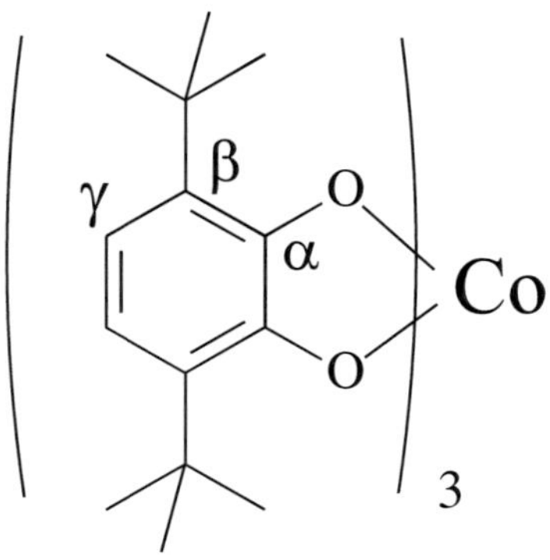

4

In compound **4**, three unpaired electrons are located on three molecules of ligands. In each of three molecules of the ligand, there are five pairs of nonequivalent carbon atoms. These are two atoms, which are directly coupled with oxygen atoms. Another pair consists of two carbon atoms, which are in $\alpha-$position relative to the two first atoms, and the third pair is formed by the carbon atoms, which are in $\beta-$position relative to the first pair. Owing to equivalence of these two carbon atoms, two hydrogen atoms coupled with them are also equivalent with each other. Therefore, one should expect that one signal from these atoms will be observed in the NMR spectrum. The fourth and fifth pairs of carbon atoms belong to the *tert*-butyl groups. The fourth pair includes carbon atoms, which are directly bonded with the cycle, whereas the fifth pair is formed by carbon atoms of the methyl groups. Protons of the methyl groups, bonded with these carbons, should also be equivalent with each other. For this reason, they should give one signal in the NMR spectrum. Notably, the ratio of signal intensities of these protons and aromatic protons should be 9:1. Therefore, it can be anticipated that ^{1}H NMR can be analyzed quite easily.

In the PMR spectrum of coordination compound **4**, which has been recorded just after dissolution in toluene-D_8, there is a high-field signal ($\delta = 1.76$ ppm) assigned to protons the *tert*-butyl groups.

It is natural to expect that another low-field signal (relative to a signal of protons of the *tert*-butyl groups) will be observed in the spectrum. Intensity of this signal should be by 9 times lower, since the ratio of the methyl and aromatic protons in compound **4** is 9:1. In reality, in the PMR spectrum, there is a signal with $\delta = 5.76$ ppm, but its intensity is significantly lesser that the intensity of a signal with $\delta = 1.76$ ppm (approximately by forty – fifty times). Moreover, apart from a signal of *tert*-butyl, another signal is detected ($\delta = 1.28$ ppm). Intensity of the latter signal is considerably lower that that of a signal of the *tert*-butyl groups, but, at the same time, it is approximately by one order more intensive in comparison with low-field resonance. Increase of the sample temperature effects slightly on positions and width of the specified resonance lines. At the same time, in the region of high-field, at a temperature ~ 70° C, a broad signal appears. This signal is shifted toward low-field and its width decreases at the temperature increase. Figure 3.15 illustrates the effect of sample temperature increase on the PMR spectra of compound **4**.

Transformation of the spectrum upon decrease of the studied sample temperature to -80° C (Figure 3.16) is of special interest. Indeed, in the low-field at ~ 0° C there is a broad signal, which first is narrowed at temperature decrease -40° C. It should be noted that the specified broad low-field signal appears in the spectra, in which high-field broad signal is absent. This fact allows these two signals to be assigned to different aromatic protons. When temperature decreases -40° C, the most intensive signal of the *tert*-butyl group protons broadens, and then is divided into two signals. Narrowinf of these two signals and the above-described signal with $\delta = 1,76$ ppm cause the appearance of another resonant signal with $\delta = 1,1$ ppm (-40° C, Figure 3.16). The second signal is significantly broader than the first one, and it is overlapped with residual signal from deuterated solvent. Further decrease of the studied sample temperature to -80° C leads to another broadening of all signals after -60° C, including low-field signal of low intensity in the spectra of the sample recorded at positive temperatures (signal $\delta = 5.45$ ppm in the spectrum shown in Fog. 2 at -80° C).

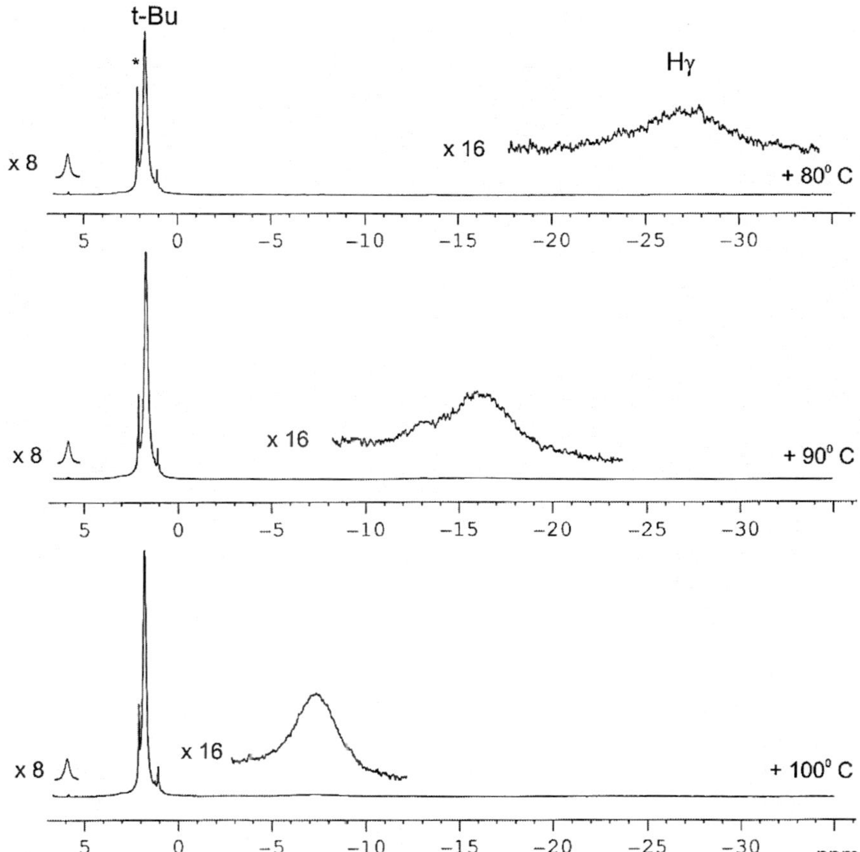

Figure 3.15. ^1H NMR spectrum of $Co(SQ)_2$ (toluene-D_8); * - the solvent signal.

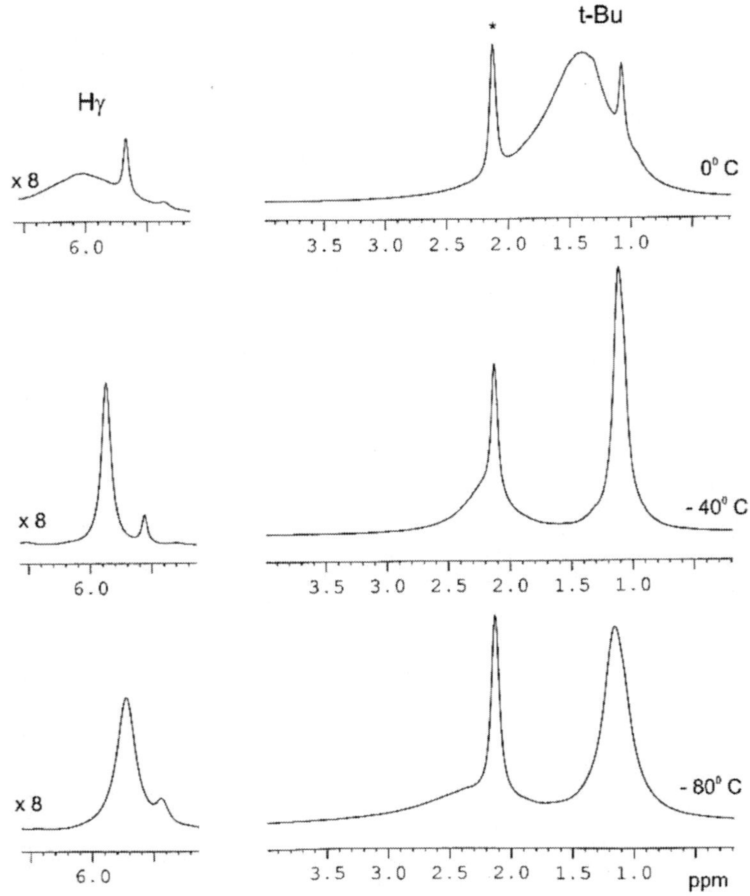

* - solvent signal.

Figure 3.16. ^{1}H NMR spectrum of $Co(SQ)_3$ (toluene-D_8).

To obtain spectral information in the conditions of high temperatures, the PMR spectra of compound **4** dissolved in DMSO-d_6 have been analyzed [299]. In the spectrum, recorded after the dissolution, a low-field signal in the region 6 – 8 ppm (δ 6,2 ppm) is assigned to H-4 and H-5 protons, whereas three overlapping signals with $\delta \approx$ are attributed to protons of *tert*-butyl groups (Figure 3.17).

When temperature of the sample increases, the low-filed signal is first narrowed (~ to 70 °C), and then begins again to broaden up to 140 °C. Decrease of temperature of the studied sample leads to narrowing and splitting of the low-field signal into two signals (~ at 90 °C), which chemical shifts in the spectrum recorded at 20 °C are 5,75 and 6,85 ppm, respectively. High-field signals also undergo noticeable transformation during variation of the sample temperature: they are narrowed and then combine in one signal (approximately at 90 °C), which also continues to be narrowed up to 140 °C. When temperature of the sample decreases in reversed order, this signal is split into two ones (after 50 °C), the values of chemical shifts being 0,83 and 1,15 ppm, respectively (40 °C) (Figure 3.18).

The spectra described here reflect intramolecular dynamics, characteristic for complex **4**. According to the current ideas about similar intramolecular processes, the observed transformation of the NMR spectra can be explained by the mechanism depicted below in Scheme LXVIII. In their turn, structures **3** can participate in further intramolecular processes (Scheme LXIX). Structures **5** and **6** are typical, since their presence is experimentally checked (at least, under certain conditions). As for paramagnetic structure **5**, it was already mentioned in the beginning of this section, HFC results in characteristic temperature dependences of paramagnetic shifts and broadenings of the NMR signals. Noteworthy, under the conditions of fast exchange, the signal shifts should obey to the Curie law, i.e. decrease upon sample temperature increase. This is also true for the signal width.

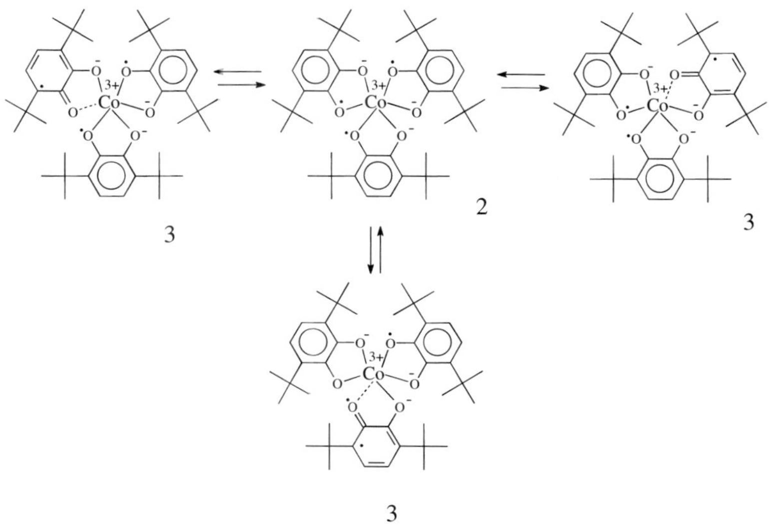

LXVIII

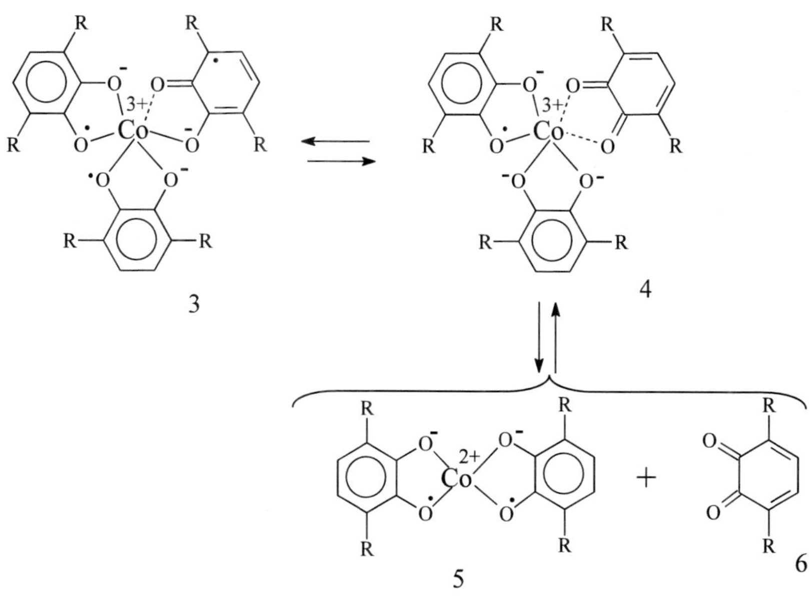

LXIX

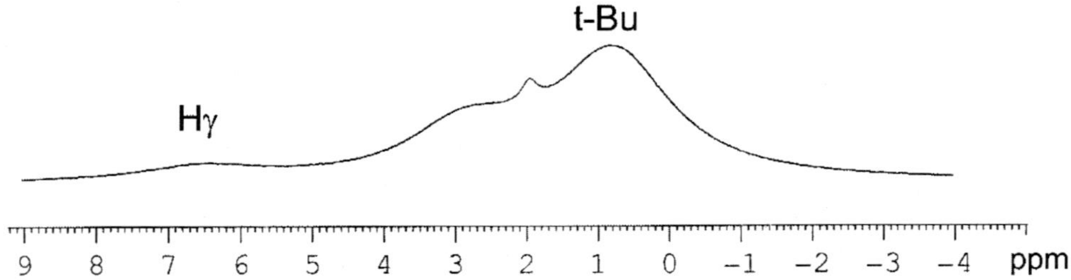

Figure 3.17.¹H NMR spectrum of $Co(SQ)_3$ (DMSO-D_6; T = 25^0 C).

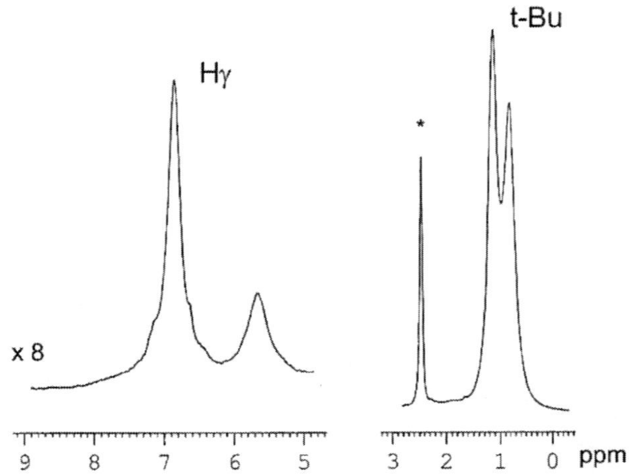

* - solvent signal.

Figure 3.18.¹H NMR spectrum of $Co(SQ)_3$ (DMSO-d_6; T = 40^0 C).

As for compound **4**, one should expect significant paramagnetic shifts and broadenings for signals of aromatic protons, but not for signals of protons of the *tert*-butyl groups. This assumption is confirmed by the NMR spectra of compound 4 dissolved in toluene-D_8 (see above). In the case of structure **6**, i.e. quinoline, which can be considered as diamagnetic analog of ligand **4**, its spectrum can be readily analyzed. The spectrum contains two singlets with $\delta = 1{,}20$ ppm (signal of the *tert*-butyl groups) and 6,17 ppm (signal of aromatic protons), which intensity ratio is 9 : 1. The δ values of these signals in the spectrum of compound **4** dissolved in DMSO-d_6 are equal 1,15 and 6,89 ppm, respectively. Another peculiarity, which should be taken into account during interpretation of the NMR spectra of compound **4**, is the hindrance of *tert*-butyl groups rotation. This effect is most probable at low temperatures, when it will significantly effect on positions and, what is more important, on broadening of signals from protons of these groups.

From the above-stated, in particular, it follows that solvents with different abilities to effect on intermolecular interactions should also influence the relative contribution of structures 2 – 6, and, therefore, the character of NMR spectra transformation in the course of variation of the sample temperature. Figure 3.19 shows temperature dependences of signal shifts of aromatic protons as well as protons of the *tert*-butyl groups in complex **4** dissolved in DMSO-d_6.

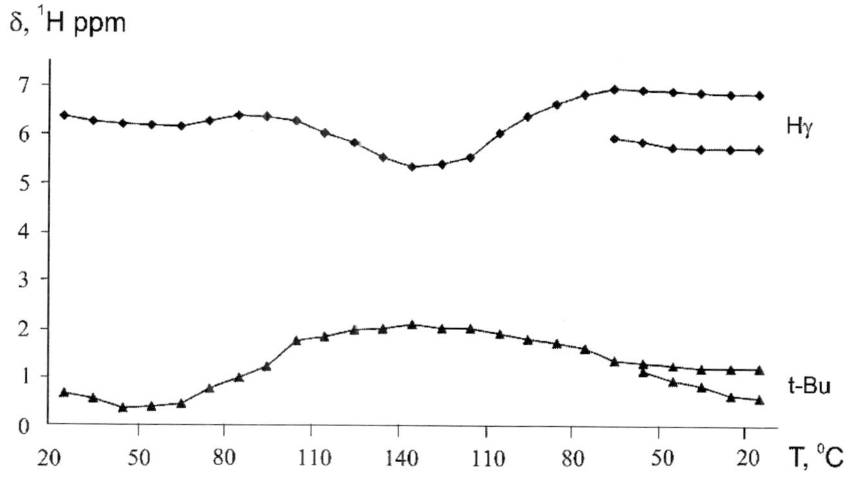

Figure 3.19. Dependence of δ values of $Co(SQ)_3$ upon temperature in DMSO– d_6 .

As is seen from Figure 3.19, the observed δ values are average. Variation of the temperature in direct and reverse direction leads to such dynamic equilibrium in the sample, when two signals of both aromatic and *tert*-butyl protons are observed in the NMR spectra (recorded at room temperature).

The NMR spectra of compound **4** dissolved in toluene-D_8 show the same picture. The only exception is an appearance of a high-field broad signal. According to the theory (chapter 1 of the monograph), paramagnetic transformation of the spectra depends both upon nature of HFC in the paramagnetic complex (which NMR spectra are detected) and conditions of the spectra recording. In particular, such transformation, and hence possibility of detecting the resonance signals is defined to a large extent by character of exchange processes with participation of paramagnetic particle. In relation to the spectra discussed in this work, the replacement of toluene-D_8 by DMSO-d_6 makes it impossible to detect the signals caused by HFC, characteristic for structure **5** (see Scheme of LXIX).

Another peculiarity observed in the NMR spectra (recorded in different solvents) of complex **4** is the aforementioned appearance of a broad signal of *tert*-butyl protons under the conditions of lower (compared to room) temperatures. This is illustrated by the spectra shown in Figure 3.16. Taking into account the temperatures, at which this effect is observed and the bulky character of *tert*-butyl substituents in the aromatic rings of compound **4**, the above difference should be attributed to steric hindrances causing chemical nonequivalence of *tert*-butyl groups.

Compounds **3** deserve a special attention, since further intramolecular exchange in complex **4** proceeds via these compounds. Besides, in these structures, the unpaired electron is localized on one of two carbon atoms (suitable for such localization) of the aromatic ring. This paramagnetic center can influence the proton signals of the *tert*-butyl substituents bonded with the specified carbon atoms. Such influence is possible, in particular, via the hyperconjugation mechanism between the orbital occupied by unpaired electron spin and the CH-bond of *tert*-butyl groups [171].

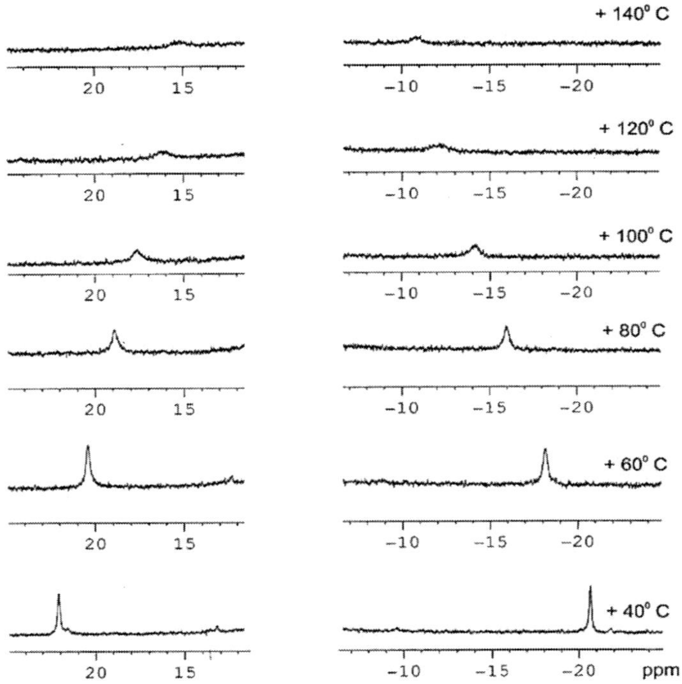

Figure 3.20. Satellite signals in the ^{1}H NMR spectra of $Co(SQ)_3$, solvent - DMSO–d_6.

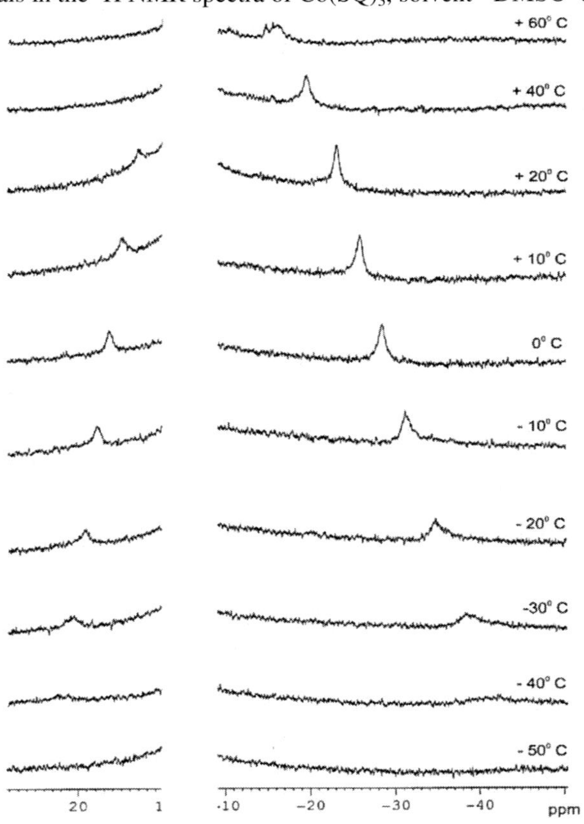

Figure 3.21. Satellite signals in the ^{1}H NMR spectra of $Co(SQ)_3$, solvent – toluene-D_8.

For further elucidation of a role of compounds 3, we should discuss here a peculiarity of compound **4** NMR spectral transformation, observed [299] in the course of the sample temperature varying. It is found that, upon heating to 110 °C (solvent DMSO-d_6), two broaden signals appear in the NMR spectra of compound **4**. One of these signals is located in higher-field as compared to the signal of *tert*-butyl protons, and the second one is detected in a low-field. These signals are preserved, when temperature of the sample decreases from 140° to room, their location and width being changed according to the Curie law.

The distance between signals changes in a similar way, increasing from 6530 Hz (26,1 ppm) at 140° C to 10700 Hz (42,8 ppm) at 40° C (Figure 3.20). In the PMR spectra of compound **4** recorded in toluene-D_8, the specified signals are also observed (Figure 3.21). They behave similarly in the course of the sample temperature variation. The distance between them changes from 9030 Hz (36,1 m ppm) at 20° C to 15900 (63,6 ppm) at -40°C. Below this temperature, the signals turn out to be strongly broadened.

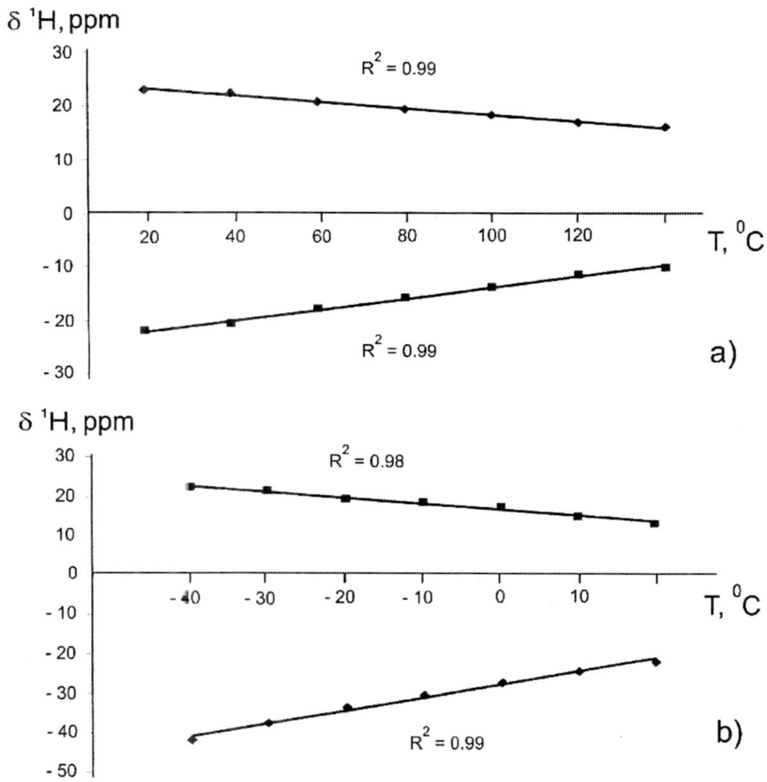

Figure 3.22. Temperature dependence of δ values of satellite signals in $Co(SQ)_3$: a) in DMSO–D_6 ; b) in toluene–D_8 .

Intensity of the aforementioned signals is significantly lower than intensity of signals from protons of the *tert*-butyl groups. This fact allows the former signals to be considered as satellite ones. It is a common knowledge that the satellite signals are quite often observed in the spectra of diamagnetic compounds with the unequal content of two and (or) more magnetic isotopes. Typical example is an appearance of satellites of methyl group signals in the PMR spectra due to spin-spin coupling between protons and a magnetic isotope ^{13}C (value of constant J_{CH} is usually about 125 Hz), which natural content is only 1,1%. Satellites reflect the nature of intramolecular magnetic interactions and therefore, for sure, do not depend on conditions of the spectrum recording (value of external magnetic field, sample temperature) if, of course, these conditions do not effect on electronic structure of the compounds.

In compound **4**, apart from carbon-13, there are two other kinds of magnetic nuclei, oxygen-17 and cobalt-59. According to general positions of the NMR theory, the PMR spectrum does not contain satellite signals owing to spin-spin coupling between protons and these nuclei. Thus, the signals detected by us should be of other nature (different from satellite signals in diamagnetic compounds). It is confirmed by temperature dependences of satellite signal shifts, which change according to the Curie law. The characteristic linear dependences are given in Figure 3.22.

From the above-stated and taking into account the discussion of spectral information obtained by us, it can be concluded that the specified satellite signals belong to a paramagnetic structure. This is a structure **3**. The presence of the latter in schemes XVIII and LXIX, reflecting the intramolecular dynamics of compounds **4**, causes appearance of signals having satellite character. Two signals correspond to two states of unpaired electron spin ($\alpha-$and$\beta-$states). One of these states is responsible for low-field shift, and the second state causes high-field shift of the *tert*-butyl proton signal relative to their positions in the spectra of diamagnetic compound. A fraction of states **3** in possible structures for compound **4** is not high, several percent in the temperature range used by us. Therefore, intensity of signals (caused by these states) appears to be essentially lesser that that of the signals from diamagnetic states.

The ^1H NMR spectra of quinoline **6** with additives of complex **4** can serve as another confirmation of schemes XVIII and LXIX validity. Indeed, the spectrum of compound **6** with additive **4** (~ 5%) shows a low-field broaden signal from aromatic protons. A signal from the *tert*-butyl protons appears to be also broad, but to a lesser extent. After heating the sample up to 140° C and cooling to room temperature, the spectrum is transformed in such a way that it contains the signals caused by protons of both compound **6** and additive **4** according to their concentration in the studied sample. Addition of complex **4** to the sample containing a solution of compound **6** in toluene–D_8, also leads to considerable broadening of a low-field signal, which then is narrowed in the course of varying the sample temperature. However, we failed to record the spectrum, which would contain the signals from **4** and **6** separately (as it was in a case of the spectra of the samples dissolved in DMSO-d_6). In other words, in all temperature range, the average spectrum is observed. This peculiarity is in agreement with the above-described characteristic features of the NMR spectrum of compound **4** dissolved in DMSO-d_6 and toluene–D_8 and can be explained by different contribution of the intermediate structures shown in Schemes LXIII and LXIX in the specified solvents.

To summarize this section, the following conclusions can be drawn.

1. ^1H NMR spectra of cobalt semi-quinolate complex recorded in a wide range of temperatures have been analyzed. It is found that the observed transformation of the spectra is caused by the intramolecular dynamics inherent in this compound.
2. A mechanism involving molecular structures, responsible for intramolecular exchange processes in the cobalt semi-quinolate complex, is proposed. A possibility

of visualization of such structures in solution by the NMR technique is shown. This allows controlling the valence tautomerism (at molecular level), which leads to alteration of valence state of the coordination ion. Thus, magnetic phase transition in the macrosample can be controlled.

3. It is found that the intramolecular exchange presupposes the obligatory presence of paramagnetic structure containing uncompensated electronic spin localized on carbon atom of the aromatic ring of the ligand molecule. In such a structure, the cobalt atom is trivalent. The presence of such structure leads to appearance of satellite signals in the NMR spectra. Apparently, this is the first experimental example of such manifestation of HFC in the conditions of intramolecular exchange.

CONCLUSION

We hope that the reader of this book will form a clear picture of a current state of the method based on application of NMR spectra transformed by electron-nuclear or hyperfine coupling and will share opinion of the authors that this method is far from to be at the end of its resources. Indeed, the combined application of HFC and other methods of spectral analysis can significantly increase their informational content. Separate usage of the obvious dependence of paramagnetic shifts and signals width upon spatial and electronic structure of paramagnetic systems allows obtaining quite often unique information on structure and dynamics of compounds at molecular level from high resolution NMR spectra.

As to the problems related to HFC, one should underline the necessity of researches aimed at expansion of our ideas about nature of electron-nuclear interactions in diverse paramagnetic complexes. It is important to continue the combined ^{1}H and ^{13}C NMR studies directed to elucidation of the nature of contact paramagnetic shifts and signal broadenings. It is known that the ^{13}C NMR spectra contain valuable information on peculiarities of molecular structure of carbon skeleton of organic compounds and their complexes, whereas signals from hydrogen nuclei, which are located on the molecular periphery, essentially complement the spectral information. HFC modeling on the basis of theoretical calculations, including ab intio quantum-chemical ones, can be a promising direction in similar research.

Another direction in investigations of HFC by high-resolution NMR spectroscopy is connected with detection of NMR signals from other nuclei, apart from ^{1}H and ^{13}C. Currently, there is a limited number of such experimental works in the literature. Moreover, the data on nature of paramagnetic shifts (especially contact ones) of NMR signals from nuclei, which valent state is defined by electrons with main quantum number more than two, are almost absent. Finally, the study of temperature dependence of paramagnetic shifts is also important. In this line, literature also contains scanty experimental data related mainly to ^{1}H and ^{13}C NMR spectra of quite simple paramagnetic complexes.

After development of the methods for detection of HFC-transformed NMR spectra, it became clear that these spectra can be useful for study of the complex formation process. This possibility was illustrated, in particular, in the present book on the example of peculiarities of molecular structure of nickel paramagnetic complexes with nitroxyl radicals as well as features intramolecular exchange and valent tautomerism in metal semi-quinolates. There are many problems which could be successfully solved by the method discussed in this, especially when using the NMR technique both on nuclei of coordinated atoms (i.e. on nuclei of heteroatoms with which ligand molecules enter the coordination scope of the central ion) and on nuclear atoms, which are the coordinating centers.

REFERENCES

[1] NMR in paramagnetic molecules: principles and applications / Eds. La Mar G. N, Horrocks W. DeW, Holm R. H. – N. Y.; L.: Acad. Press, 1973.-519 p.

[2] Nuclear magnetic resonance shift reagent / Ed. Sievers R. E. – N. Y.; L.: Acad. Press, 1973, 410 p.

[3] Voronov, VK. Method of paramagnetic additives in NMR spectroscopy, *Novosibirsk: Nauka*, 1989, 168 p.

[4] Knight, WD. Electron paramagnetism and nuclear magnetic resonance in metals. *Solid State Phys,* 1956, Vol. 2, 93-136.

[5] Eaton, DR; Phillips, WD. Nuclear magnetic resonance in paramagnetic compounds. *Zh. Strukt. Khim,* 1968, Vol. 9, № 1, 153-183.

[6] Buchachenko, AL. Complexes of radicals and molecular oxygen with organic molecules, M.: Nauka, 1984, 157 p.

[7] Hinckley, CC. Paramagnetic shifts in solutions of cholesterol and the dipyridine adduct of tris(dipivalomethanato) europium (I I I). *J. Am. Chem. Soc,* 1969, Vol. 91, N 18, 5160-5162.

[8] Sanders, JKM; Williams, DH. A shift reagent for use in NMR spectroscopy. A first-order spectrum of *n*-hexanol. *Chem. Com,* 1970, N 7, 422-423.

[9] Eaton, DR; Josev, AD; Phyllips, WD; Benson, RE. Hyperfine contact shifts and spin density distribution in some nickel(I I) chelates. *Dis. Far. Soc,* 1962, N 34, 77-87.

[10] Happe, JA; Ward, RL. Isotropic NMR shift pyridine-type bases complexed with paramagnetic nickel(I I) and cobalt(I I) acetylacetonates. *J. Chem. Phys,* 1963, Vol. 39, N 5, 1211-1218.

[11] Chuvylkin, ND; Zhidomirov, GM. Ab initio quantum-chemical study of σ-electron system and calculations of σ-electron radicals. *Zh. Strukt. Khim,* 1971, Vol. 12, № 2, 347-376.

[12] Doddrell, D; Roberts, JD. Observation of carbon-13 resonance in a paramagnetic transition metal complex. Nickel (I I) N, N´-di(p-tolyl) aminotroponominate. *J. Am. Chem. Soc,* 1970, Vol. 92, N 14, 4484-4486.

[13] McConnell, HM. Spin density matrices for paramagnetic molecules. *J. Chem. Phys,* 1958, Vol. 28, N 6, 1188-1192.

[14] Karplus, M; Fraenkel, GK. Theoretical interpretation of carbon-13hyperfine interactions in electron spin resonance spectra. *J. Chem. Phys,* 1961, Vol. 35, N 4, 1312-1323.

[15] Doddrell, D; Roberts, JR. The use of carbon-13 isotropic shifts as a probe of mechanisms of spin-transmission in metal complexes. Some pyridine-type bases and triphenylphosphine complexed to nickel (I I) and cobalt(I I) acetylacetonates. *J. Am. Chem. Soc,* 1970, Vol. 92, N 23, 6839-6844.

[16] Sagdeev, RZ; Molin, YuI; Dvornikov, EV; et al. Hyperfine coupling with ^{13}C nuclei in paramagnetic complex compounds. *Zh. Strukt. Khim,* 1971, Vol. 12, № 2, C. 245-251.

[17] Zaev, EE; Molin, YuN; Zhidomirov, GM; Voevodsky, VV. Correlation between regularities of spin density delocalization: ESR study of radicals and NMR spectra of molecules and complexes of paramagnetic ions. *Dokl. Akad. Nauk USSR,* 1967, № 6, 1370-1373.

[18] Horrocks, W DeW; Jonston, DL. Spin delocalization in systems. *Inorg. Chem,* 1971, Vol. 10, N 8, 1835-1838.

[19] Cramer, RE; Drago, RS. Nuclear magnetic resonance contact shifts and delocalization mechanisms in octahedral nickel(I I) complexes. *J. Am. Chem. Soc,* 1970, Vol. 92, N 1, 66-70.

[20] De Simone, RE; Drago, RS. On the nuclear magnetic resonance spectra of tris(bipyridyl) iron (I I I). *Inorg. Chem,* 1972, Vol. 11, N 3, 668-670.

[21] Scarlett, MJ; Casey, AT; Craig, RA. INDO calculations of spin density. Aust. J. Chem, 1970. Vol. 23, N 7, 1333-1344.

[22] Horrocks, W DeW. Spin delocalization in σ-systems. An analysis of INDO calculations. *Inorg. Chem,* 1973, Vol. 12. N 5, 1211-1212.

[23] Cramer, RE; Chudyk, MA. Nuclear magnetic resonance contact shifts and delocalization mechanisms in tris(acetylacetone) nickel(I I) perchlorate. *J. Magn. Res,* 1973, Vol. 12, N 2, 168-181.

[24] Zaev, EE; Molin, YuN. Paramagnetic shifts in NMR spectra and structure of Ni(II) and Co(II) acetylacetonate complexes with pyridine-N-oxide. *Zh. Strukt. Khim,* 1966, Vol. 7, № 5, 680-685.

[25] Horrocks, W DeW; Hall, DD. Direct evaluation of dipolar nuclear magnetic resonance shifts from single-crystal magnetic measurements. Paramagnetic anisotropy of bis(2,4-pentanedionate) bis(pyridine) cobalt(I I). *Inorg. Chem,* 1971, Vol. 10, N 10, 2368-2370.

[26] Horrocks, W DeW. Orbital energies as ligand field parameters in the weak-field approximation. Application to the paramagnetic anisotropy of bis(2,4-pentanedinato)bis(pyridine) cobalt(I I). *Inorg. Chem,* 1974, Vol. 13, N 11, 1775-2778.

[27] Horrocks, W DeW. A new general approach to ligand field calculations. The effective perturbation method. *J. Am. Chem. Soc,* 1972, Vol. 94, N 2, 656-658.

[28] Berus, EN; Gladky, YuG; Barkhash, VA; Molin, YuN. Effect of substituents on covalence degree of metal-ligand bond in complexes of substituted anilines with nickel acetylacetonate. *Dokl. Akad. Nauk USSR,* 1971, Vol. 197, № 6, C. 1362-1365.

[29] Berus, EN; Schastnev, PV; Barkhash, VA; Molin, YuN. NMR study of effect of ligand and chelate nature on covalence degree of metal-ligand bond in complexes of paramagnetic metals. *Zh. Strukt. Khim,* 1973, Vol. 14, № 1, 34-43.

[30] Mel'nikova, NV; Pilipenko, AG. Paramagnetic shifts in PMR spectra of pyridine, α- and β-picoline coordinated with nickel chelates. *Coord. Khim,* 1975, Vol. 1, № 12, 1703-1705.

[31] Obynochny, AA; Sagdeev, RZ; Molin, YuN; Rezvukhin, AI. Stereospecific regularities of hyperfine coupling in complexes of $Ni(AA)_2$ with cyclic and linear amines. *Coord. Khim,* 1975. Vol. 1, № 6, 817-823.

[32] Obynochny, AA; Sagdeev, RZ; Molin, YuN; Rezvukhin, AI. ^{13}C NMR spectra of Ni(II) complexes with pyridines. Izv. SO Akad. Nauk USSR. *Ser. Khim,* 1975, № 14, 114-119.

[33] Voronov, VK. Paramagnetic shifts in nickel complexes with some derivatives of five-membered N-heterocycles. Izv. SO Akad. Nauk USSR. *Ser. Khim.* 1971, № 4, 62-65.

[34] Anstey, P; Orrell, KC. NMR studies of nickel(I I) complexes and theoretical treatment of spin-density distributions in cobalt(I I) and nickel(I I) complexes. *J. Chem. Soc. Dalt,* 1974, N 16, 1711-1717.

[35] Voronov, VK; Moskovskaya, TE; Glukhikh, VI; et al. NMR study of paramagnetic complexes of iron group elements with 1-vinylimidazoles. *Coord. Khim,* 1979, Vol. 5, № 2, 218-227.

[36] Wicholas, M; Mustacich. R; Johnston, B; et al. Proton nuclear magnetic resonance contact shifts of octahedral iron(I I), cobalt(I I) and nickel(I I) imidazole complexes. *J. Am. Chem. Soc,* 1975, Vol. 97, N 8, 2113-2117.

[37] Morishima, J; Okade, K; Jonezawa, T. NMR contact shifts and electron spin distribution: Proton and carbon-13 contact shift studies of azanaphthalenes. *J. Am. Chem. Soc,* 1972, Vol. 94, N 5, 1425-1430.

[38] Dhingra, MM; Govel, G. Kanekar, CR. NMR contact shifts of spin-delocalization in adducts of nickel(I I) diethyldithiophosphate with secondary amines. *J. Magn. Res,* 1972, Vol. 6, N 4, 577-585.

[39] Hayamizu, K; Murata, M; Yamamoto, O. The ^{13}C magnetic resonance study on the effects of divalent paramagnetic metal acetylacetonates on pyridine, quinoline and isoquinoline. Specific interactions with $Mn(AA)_2$. *Bul. Chem. Soc. J,* 1975, Vol. 48, N 6, P, 1842-1847.

[40] Wierseman, RJ; Hawthorme, MF. Electron delocalization in paramagnetic metallocarboranes. *J. Am. Chem. Soc,* 1974, Vol. 96, N 3, 761-770.

[41] Nivorozhkin, LE; Konstantinovsky, LE; Minkin, VI; et al, Contact shifts and distribution of spin density in paramagnetic nickel bis- (5-thiopyrazolones), *Zh. Inorg. Khim,* 1975, Vol. 20, № 11, 3012-3016.

[42] Eaton, DR. The NMR of some paramagnetic transition metal acetylacetonates. *J. Am. Chem. Soc,* 1965, Vol. 87, N 14, 3097-3102.

[43] Johnson, BFG; Levis, J; McArdler, P; Horton, JR. Contact effects in NMR spectra with lanthanide shift reagents. *Chem. Com,* 1972, N 9, 535.

[44] Ajsaka, K; Kainoaho, M; Shigemoto, H; et al. Conspicuous effects due to complex formation and contact term upon lanthanide-induced shift in ^{19}F NMR spectra of some fluoroaromatic. *Chem. Let,* 1973, N 7, 1205-1208.

[45] Tori, K; Yoshimura, Y; Kainoshe, M; Ajisaka, K. Evidence for the presence of contact term contribution to lanthanide induced shifts in ^{1}H and ^{13}C NMR spectra of pyridine N-oxides. *Tetrahedron Lett,* 1973, N 17, 1573-1576.

[46] Golding, RM; Halton, MP. A theretical study of the ^{14}N and ^{17}O NMR shifts in lanthanide complexes. *Aust. J. Chem,* 1972, Vol. 25, N 12, 1577-1581.

[47] Bleany, B; Dobson, CM; Levine, BA; et al. Origin of lanthanide NMR shifts and their uses. *Chem. Com,* 1972, N 13, 791-793.

[48] Reuben, J. On the origin of chemical shifts in lanthanide complexes and some implications therefore. *J. Magn. Res,* 1973, Vol. 11, N 1, 103-104.

[49] Gansow, OA; Loeffler, PA; Davis, RE; et al. Evalution of lanthanide-induced carbon-13 contact vs. pseudocontact NMR shifts. *J. Am. Chem. Soc,* 1973, Vol. 95, N 10, 3389-3390.

[50] Beech, G; Morgan, RI. Anomaleous substituent effects observed for lanthanide-induced shifts in substituted pyridines. *Tetrahedron Lett,* 1974, N. 12, 973-976.

[51] Gansow, OA; Loeffler, PA; Davis, RE; et al, Evidence for europium-induced Fermi contact shifts in carbon-13 NMR spectra. *J. Am. Chem. Soc,* 1973, Vol. 95, N 10, 3390-3392.

[52] Hawkes, GE; Marzin, C; Johns, SR; Roberts, JD. NMR shift reagents. Quantitative estimates of contact contributions to lanthanide-induced chemical-shift changes for exo-norbornylamine. *J. Am. Chem. Soc,* 1973, Vol. 95, N 5, 1661-1662.

[53] Cramer, RE; Davis, R; Seff, K. Calculation of lanthanide-induced shifts from molecular structure. *J. Am. Chem. Soc,* 1974, Vol. 96, N 13, 4125-4131.

[54] Johnson, BF; Lewis, J; McArdler, JR; Norton, JR. Applications of lanthanoid shift reagents to ^{13}C and ^{1}H NMR spectroscopy. *J. Chem. Soc. Dalt,* 1974, N 12, 1253-1258.

[55] Hirayama, M; Sato, M. Quantitative estimates of contact term contributions to the lanthanide-induced ^{1}H and ^{13}C shifts in a aniline and *p*-toluidine. *Chem. Lett,* 1974. N 6, 725-728.

[56] Davis, RE; Willcott, MR. Interpretation of the pseudo-contact model for NMR shift reagents. II. Significance testing of agreement factor R. *J. Am. Chem. Soc,* 1972, Vol. 94, N 5, 1744-1745.

[57] McConnell, NM; Robertson, RE. Isotropic nuclear resonance shifts. *J. Chem. Phys,* 1958, Vol. 29, N 6, 1361-1365.

[58] Dobson, CM; Williams, RJP; Xavier, AV. Separation of contact and pseudo-contact contributions to shifts induced by lanthanide(III) ions in NMR spectra. *J. Chem. Soc. Dalt,* 1973, N 23, 2662-2664.

[59] Chalmers, AA; Pachler, KGR. Gd(FOD)$_3$-induced contact shifts. *J. Chem. Soc/ Perkin I I,* 1974, N 7, 748-752.

[60] Ajisaka, K; Kainosho, M. Contact and pseudo-contact lanthanide-induced shifts in the nuclear magnetic resonance spectrum of quinoline. *J. Am. Chem. Soc,* 1975, V. 97, N 2, 330-334.

[61] Huber, H; Pascual, C. Uber der Linfluss des Kontaktterms auf Lanthaniden-verschiebung der NMR Signale aromatischer Stickestoffheterocyclen. *Helv. Chim. Acta,* 1971, Vol. 54, N 3, 913-915.

[62] Radeglia, R. Zur Unterscheidung von Kontakt- und pseudo-kontakt-wechselwirkungsbeiträgen von Vinylogen N, N-dimethylformamiden. *Z. Chem,* 1974, Vol. 14, N 10, 417-418.

[63] Hirayama, M; Edagava, E; Hanyu, Y. Contact term contribution to lanthanide-induced ^{13}C NMR shifts in pyridine and α -picoline. *Chem. Com,* 1972, N 24, 1342-1344.

[64] Sagdeev, RZ; Obynochny, AA; Molin, YuN; Rezvukhin, AI. NMR study of paramagnetic shifts nature in Fe(II) complexes with nitrogen-containing ligands. Izv. SO Akad. Nauk USSR. *Ser. Khim,* 1975, № 12, 106-110.

[65] Hinkley, CC. Applications of rare earth nuclear magnetic resonance Shift reagents. II. The assignment of the methyl proton magnetic resonances of *d*-camphor. *J. Org. Chem,* 1970, Vol. 35, N 8, 2834-2836.

[66] Rondeau, RE; Sievers, RE. New nuclear magnetic resonance shift reagents. *Anal. Chem,* 1973, Vol. 45, N 12, 2145-2147.

[67] Beyer, K. Tris-(4,4,4-trifluoro-1-(2-thienyl)-1,3-butandiono) europium(III) as NMR-verschiebungsreagens. *OMR,* 1973, Vol. 5, N 10, 471-472.

[68] Potapov, VM; Rukhadze, EGIlyina, IG. Bakhmutskaya VG, Europium and praseodymium tenoyltrifluoroacetonates as shifting reagents in nuclear magnetic resonance. *Zh. Org. Khim.* 1974, Vol. 44, № 2, 462-463.

[69] Dushkin, AV; Zolin, VF; Koreneva, LG; et al. New shift-reagents for study of molecular structures using NMR methods in polar solvents. *Zh. Strukt. Khim,* 1974, Vol. 15, № 3, 410-413.

[70] Weiner, LM; Zolin, VFKoreneva, LG; Sagdeev, RZ. A possibility of some europium complexes application as shifting reagents. *Coord. Khim,* 1975, Vol. 1, № 11, 1512-1515.

[71] Morrill, TC; Clark, RA; Bilobran, D; Joungs, DS. New dimensions in lanthanide shift reagent-PRM analysis of organic compounds: $Eu(tfh)_3$. *Tetrahedron Lett,* 1975, N 6, 397-398.

[72] Potapov, VM; Bakhmutskaya, VG; Ilyina, IG; et al. Formylcyclohexanone-based lanthanide shifting reagents for nuclear magnetic resonance. *Zh. Org. Khim,* 1976, Vol. 46, № 9, 2117-2121.

[73] Johnston, MF; Forsberg, JH. Synthesis and spectroscopic studies of β,β',β''-triaminotriethylamine chelates of trivalent lanthanide perchlorates. *Inorg. Chem,* 1976, Vol. 15, N 3, 734-737.

[74] Maskasky, JR; Kenney, ME. Macrocyclic nuclear magnetic resonance shift reagents. *J. Am. Chem. Soc,* 1973, Vol. 93, N 5, 1443-1448.

[75] Mooney, JR; Choy, CK; Knox, K; Kenney, ME. Determination of Si, O, Si bond angle common to the shift reagent compounds $(CH)_3SiO(P_cSiO)_xSi(CH_3)_3$ by an induced shift technique. *J. Am. Chem. Soc,* 1975, Vol. 97, N 11, 3033-3038.

[76] Folcher, G; Paris, J; Plurien, P; et al. A novel and efficient NMR shift reagent: tetrakis-(1,1,1-trifluoro-4-phenylbutane-2,4-dionato) uranium. *Chem. Com,* 1974, N 1, 3-4.

[77] Boze, AK; Siguira, M; Srinivasen, PR. Arsenic trichloride as a convent solvent for C-13 NMR spectroscopy. *Tetrahedron Lett,* 1975, N 14, 1251-1254.

[78] Evans, DF; Tucker, JN; de Villard, GC. Lanthanide shift, reagents for alkenes. *Chem. Com,* 1975, N 6, 205-206.

[79] Wenzel, TJ; Lalonde, DR. New binuclear NMR shift reagent for olefin and aromatics. *J. Org. Chem,* 1983, Vol. 48, 1951-1954.

[80] Dambska, An/, Janowski, A. The origin of shifts induced complex silver lanthanide shift reagents in ^1H NMR spectra of unsaturated hydrocarbons. *J. Magn. Reson,* 1984, Vol. 59, 13-19.

[81] Raber, DJ; Hajek, M. Binuclear shift reagents structure and mode of interaction of Ag(fod), $Zn(fod)_3$ with benzene. Magn. Reson. Chem, 1986, Vol. 24, 297-303.

[82] Boze, AKSrinivasan, PR; Trainor, G. Titanium tetrachloride as a shift reagent. *J. Am. Chem. Soc,* 1974, Vol. 96, N 11, 3670-3671.

[83] Boze, AK; Srinivasan, PR. Titanium tetrachloride-induced shift in carbon-13 NMR spectra. *J. Magn. Res,* 1974, Vol. 15, N 3, 592-593.
[84] Sarneski, JE; Reilley, CH. NMR studies of configuration and ligand conformation in paramagnetic octahedral complexes of nickel(I I). *Inorg. Chem,* 1974, Vol. 13, N 4, 977-988.
[85] Stone, EW; Maki, AH. Hindered internal rotation and ESR spectroscopy. *J. Chem. Phys,* 1962, Vol. 37, N 6, 1326-1333.
[86] Adam, FC; King, FW. Angular dependence of the β-proton isotropic hyperfine coupling constant. *J. Chem. Phys,* 1973, Vol. 58, N 6, 2446-2453.
[87] Stock, LM; Wasielewski, MR. Pseudocontact and contact shifts for 6-aminobensobicyclo(2,2,2)octane and 2-aminotriptycene. The sign of B_0. *J. Am. Che. Soc,* 1973, Vol. 95, N 8, 2743-2744.
[88] Maruani, J. Dependances conformationnelles des couplages hyper-fins de noyaux en position dans les radicaux organiques. *Mol. Phys,* 1975, Vol. 30, N 6, 1685-1691.
[89] Stock, LM; Wasielewski, MR, Angular dependence of the β-carbon atom hyperfine coupling constant. *J. Am. Chem. Soc,* 1974, Vol. 96, N 2, 583-585.
[90] Knorr, R; Polses, H; Bischer, E. Conformation analysis by spin transmission into rotating alkyl groups. *J. Am. Chem. Soc,* 1975, Vol. 97, N 3, 643-644.
[91] Morishima, I; Okada, K; Okashi, M; Yonezawa, T. Conformation dependence of the proton contact shifts for σ-electron system: a 220 MHz NMR study. *J. Che. Soc. D,* 1971, N 1, 33-34.
[92] Morishima, I; Yoshikawa, K, Homoallyl interaction between the nitrogen lone pair and the nonadjacent σ-bond. *J. Am. Chem. Soc,* 1975, Vol. 97, N 11, 2950-2956.
[93] Nivorozhkin, LE; Nekipelov, VM; Konstantinovsky, LE; et al. Configurational inversion in tetrahedral nickel (II) bis-(5-thio(seleno)-pyrazole-4-aldimines. *Zh. Org. Khim,* 1985, Vol. 55, № 4, 849-861.
[94] Mastikhin, VM; Nekipelov, VM; Zamarayev, KI. New possibilities of NMR spectroscopy in catalytic studies. *Kinet. Catal,* 1982, Vol. 23, № 6, 1323-1333.
[95] Zaev, EE; Voronov, VK; Shwartsberg, MS; et al. Application of paramagnetic additions to the structure determination of some pyrazoles by NMR. *Tetrahedron Lett,* 1968, N 5, 617-622.
[96] Natusch, DFS. Conformational determination in paramagnetic metal complexes. *J. Am. Chem. Soc,* 1973, Vol. 95, N 5, P.1688-1690.
[97] Doddrell, D; Burfitt, I; Piggs, NV. Use of paramagnetic broadening agent to assign the ^{13}C NMR spectra of some aliphatic amins. *Aust. J. Chem,* 1975, Vol. 28, N 2, 369-379.
[98] Levy, GC; Komoroski, RA. Paramagnetic relaxation reagents. *J. Am. Chem. Soc,* 1974, Vol. 96, N 3, 678-681.
[99] Led, JJ; Crant, DM. Structural and dynamical information on a manganese (II) complex of histidine. *J. Am. Chem. Soc,* 1975, Vol. 97, N 24, 6962-6970.
[100] Lindoy, LF. NMR studies of coordination metal complexes using lanthanide shift reagent. *Coord. Chem. Rev,* 1983, Vol. 48, N 1, 83-100.
[101] Shue, FF; Yen, TF. Application of lanthanide NMR shift reagents for functional group studies of shale bitumen asphaltenos. *Fuel,* 1983, Vol. 62, 127-128.
[102] Kalabin, GA; Kanitskaya, LV; Kushnarev, DV. Quantitative NMR spectroscopy of natural organic feedstock and products of its processing, M.: Khimiya, 2000, 408 p.

[103] Honeybourne, CL. Concerning the location of the principal magnetic axis in lanthanide shift reagent adducts. *Tetrahedron Lett,* N 12, 1095-1098.

[104] Horrocks, W DeW; Sipe, JP. Paramagnetic anisotropy of shift reagent adducts. Science, Vol. 177, 994-996.

[105] Wing, RM; Uebel, JJ; Andersen, KK. Lanthanide induced NMR shifts. A structural and computational study. *J. Am. Chem. Soc,* 1973. Vol. 95, N 18, 6046-6056.

[106] Newman, RH. The importance of non-axial symmetry in the interpretation of lanthanide-induced shifts for ketones. *Tetrahedron*, 1974, Vol. 30, N 8, 969-975.

[107] Voronov, VK; Keiko, VV; Moscovskaya, TE. Paramagnetic reagents for NMR study of heteroatomic compounds structure. *Zh. Strukt. Khim,* 1977, Vol. 18, № 5, 917-952.

[108] Horrocks, W DeW. Model with accounts for the apparent axial symmetry of shift reagent adducts in solution. *J. Am. Chem. Soc,* 1974, Vol. 96, N 9, 3022-3024.

[109] La Mar, GN; Metz, EA. Dipolar relaxation in shift reagents as a solution structural probe. *J. Am. Chem. Soc,* 1974, Vol. 96, N 17, 5611-5613.

[110] La Mar, GN; Faller, JW. Strategies for the study of structure using lanthanide reagents. *J. Am. Chem. Soc,* 1973. Vol. 95, N 11, 3817-3818.

[111] Faller, JW; Adams, MA; La Mar, GN. The application of lanthanide relaxation reagents in CMR. *Tetrahedron Lett,* 1974, N 9, 699-702.

[112] Faller, JW; La Mar, GN. Chemical spin-decoupling and lanthanide reagents. *Tetrahedron Lett,* 1973, N 16, 1381-1384.

[113] Caton, GA; Hart, FA; Moss, GP. Studies of the conformation of the NMR shift reagent tris(2,6,6-tetramethylheptane-3,5-dinato)europium(III) and its adducts by means of fluorescence spectra. *J. Chem. Doc. Dalt,* 1975, N 3, 221-226.

[114] Cramer, RE; Dubois, CF; Furuike, CF. Calculations of lanthanide-induced shift from molecular structure. *Inorg. Chem,* 1975, Vol. 14, N 5, 1105-1107.

[115] Johnston, MD; Shapiro, BL; Shapiro, ML. Lanthanide-induced shifts in proton nuclear magnetic resonance spectra. XI. Equilibrium constants and bound shifts for cyclohexanones and cyclohexanols. *J. Am. Chem. Soc,* 1975, Vol. 97, N 3, 542-554.

[116] Andersen, NH; Bottino, BJ; Moore, A; Shaw, JR. Metal ion probes of molecular geometry. *J. Am. Chem. Soc,* 1974, Vol. 96, N 2, 603-604.

[117] Voronov, VK. Peculiarities of lanthanides chelate application in NMR spectroscopy of weak Lewis bases. Izv. Akad. Nauk USSR. *Ser. Khim,* 1976, № 9, 2110-2112.

[118] Voronov, VK; Garnovsky, DA; Pertsikov, BZ; et al. ^{13}C and ^{1}H NMR study of complex formation of 1-methyl-2-aminobenzimidazole with Co(II) and Zn(II) *Coord. Khim.* 1986. Vol. 12, № 9, 1244-1247.

[119] Banci, L; Benelli, C; Gatteschi, D. Magnetic resonance spectra of a heterodinuclear (Ni-Co) metal complex. *Inorg. Chem,* 1981, Vol. 20, N 12, 4397-4400.

[120] Köhler, FH; Geike, WA ^{29}Si and ^{119}Sn NMR data from paramagnetic metallocenes. *J. Magn. Res,* 1983. Vol. 53, N 2, P. 297-302.

[121] Eaton, DR; McClellan, WR. d – p-Bonding and conjugation involving group IV elements. *Inorg. Chem,* 1967, Vol. 6, N 12, 2134-2138.

[122] Hirayama, M; Kitami, K. Observation of ^{57}Al NMR paramagnetic shifts (large contact shift contributions) induced by lanthanide shift reagents in tris(acetylacetonato) aluminium(III). *Chem. Com,* 1980, N. 15, 1030-1031.

[123] Bleany, B; Dobson, CM; Levine, BA; et al. Origin of lanthanide NMR shifts and their uses. *Chem. Com,* 1972, N 13, 791-793.

[124] Pinkerton, AA; Rossier, M; Spiliadis, S. Lanthanide-induced contact shifts. The average electron spin polarization, theory and experiment. *J. Magn. Res,* 1985, Vol. 64, N 3, 420-425.

[125] Fermi, E. Über die magnetischen Momente der Atomkerne. *Z. Phys,* 1930, Bd 60, S. 320-333.

[126] Nordenskiold, L; Laaksonen, A; Kowalewski, I. Applicability of the Solomon-Blombergen equation to the study of paramagnetic transition-metal-water complexes. An ab initio SCF-MO study. *J. Am. Chem. Soc,* 1982, Vol. 104, N 2, 379-382.

[127] Sunil, KK; Harrison, IF; Roger, MT. An SCF-MS-X study of a series of *d*-transition metal oxohalo complexes. *J. Chem. Phys,* 1982. Vol. 76, N 6, 3087-3097.

[128] Zhidomirov, GM; Schastnev, PV; Chuvylkin, ND. Quantum-chemical calculations of magnetic-resonance parameters, *Novosibirsk: Nauka,* 1978, 367 p.

[129] Plakhutin, BN. Quantum-chemical calculations of spin-hamiltonian constants in complexes of transition metals: Ph.D. Thesis, *Novosibirsk,* 1983, 161 p.

[130] Danovich, DK; Plakhutin, BN; Voronov, VK. Quantum-chemical study pf structure and isotropic chemical shifts in NMR spectrum of paramagnetic complex of 1-vinylimidazole with $CuCl_2$. *Theor. Experiment. Khim,* 1990, Vol. 26, № 2, 225-229.

[131] Kurland, RJ; McGarvey, BR. Isotropic NMR shifts in transition metal complexes: the calculation of Fermi contact and pseudocontact terms. *J. Magn. Res,* 1970, Vol. 2, N 3, 286-301.

[132] Bleany, B. NMR shifts in solution due to lanthanide ions. *J. Magn. Res,* 1972, Vol. 6, N 1, 91-100.

[133] Eicher, H. Theory of NMR shifts and ENDOR spectroscopy in paramagnetic complexes. *J. Chem. Phys,* 1984, Vol. 86, N 3, 331-338.

[134] Jesson, JP. Teory of isotropic nuclear resonance shifts in octahedral Co^{2+} systems. *J. Chem. Phys,* 1967, Vol. 47, N 2. 579-581.

[135] Golding, RM; Pyykkö. On the theory of pseudocontact NMR shift due to lanthanide complexes. *Mol. Phys,* 1973, Vol. 26, N 6, 1389-1396.

[136] Golding, RM; Pascual, RO; McGarvey, BR, An evaluation of the accuracy of calculating hyperfine interactions for *f*-electrons using the "point dipole approximation" *J. Magn. Res,* 1982, Vol. 46, N 1, 30-42.

[137] Voronov, VK. On the nature of paramagnetic shifts indused by lanthanide shifting reagentsIzv. Akad. Nauk USSR. *Ser. Khim,* 1977, № 2, 422-424.

[138] Lipovchenko, EL. On shielding of hyperfine coupling in calculation of pseudo-contact shifts of the nuclear magnetic resonance in paramagnetic systems. *Izv. Vuzov. Physics,* 1987. № 11, 53.

[139] Horrocks, W DeW. Evaluation of dipolar nuclear magnetic resonance shifts. *Inorg. Chem,* 1970, Vol. 9, N 3, 690-692.

[140] Horrocks, W DeW. The temperature dependencies of lanthanide-induced NMR shifts: Evaluation of theoretical approaches and experimental evidence. *J. Magn. Res,* 1977, Vol. 26, N 2, 333-339.

[141] Stout, EW; Gutowsky, HS. On the temperature dependence of lanthanide-induced NMR shifts. *J. Magn. Res,* 1976, Vol. 24, N 3, 389-398.

[142] McGarvey, BR. Temperature dependence of the pseudocontact shift in lanthanide shift reagents. *J. Magn. Res,* 1979, Vol. 33, N 3, 445-455.

[143] Slichter, Ch. Principles of magnetic resonance, M.: Mir, 1981, 218. (In Russian).

[144] Solomon, J. Relaxation processes in a system of two spins. *Phys. Rev,* 1955, Vol. 99, N 2, 559-565.

[145] Bloembergen, N. Proton relaxation times in paramagnetic solutions. *J. Chem. Phys,* 1957, Vol. 27, N 2, 595-596.

[146] Sternlicht, H. Nuclear relaxation induced by paramagnetic ions having anisotropic g-factors. *J. Chem. Phys,* 1965, Vol. 42, N 6, 2250-2251.

[147] Rubinstein, M; Baram, A; Luz, Z. Electronic and nuclear relaxation solutions of transition metal ions with spin S = 3/2 and 5/2. *Mol. Phys,* 1971, Vol. 20, N 1, 67-80.

[148] Woessner, DE. Nuclear spin relaxation in ellipsoids undergoing relational Brownioen motion. *J. Che. Phys,* 1962, Vol. 37, N 3, 647-654.

[149] Woessner, DE. Spin relaxation processes in a two-proton system undergoing anisotropic reorientation. *J. Chem. Phys,* 1962, Vol. 36, N 1, 1-4.

[150] Doddrell, DM; Healy, PC; Bendall, MR. Ligandcentered ^{1}H and ^{13}C electron-nuclear dipolar relaxation in Ni(I I) complex. Breakdown of the Solomon-Bloembergen equation. *J. Magn. Res,* 1978, Vol. 29, N 1, 163-166.

[151] Burlamacchi, L; Monduzzi, M; Saba, I. NMR, EPR and INDO studies on the complexes of dopamine with Cu^{2+}, Mn^{2+} and Fe^{2+} in aqeous solution. *J. Magn. Res,* Vol. 55, N 1, 39-50.

[152] Golding, RM; Stubbls, LC. NMR shifts in paramagnetic system: a nonmultipole expansion method. *J. Magn. Res,* 1979, Vol. 33, N 3, 627-647.

[153] Bertini, I; Luchinat, C; Kowalewski, J. Nuclear spin relaxation in paramagnetic (S > ½) systems. A comparison of two new theoretical approaches. *J. Magn. Res,* 1985, Vol. 62, N 2, 235-241.

[154] Eaton, DR; Iosey, AD; Phillips, WD; Benson, RE. Proton spin-spin coupling constants from the NMR spectra paramagnetic molecules. *J. Chem. Phys,* 1963, Vol. 39, N 12, 3513-3518.

[155] Voronov, VK; Ushakov, IA; Baikalova, LV. NMR spectra of 1-vinylimidazole paramagnetic complexes with of iron group elements. Izv. Akad. Nauk. *Ser. Khim,* 2005, № 6, 1430-1433.

[156] Kluiber, RW; Horrocks, W DeW. Proton magnetic resonance studies of paramagnetic quinoline 1-oxide and isoqunoline 2-oxide complexes. *J. Am. Chem. Soc,* 1966, Vol. 88, N 7, 1399-1404.

[157] McConnell, NM. Reaction rates by nuclear magnetic resonance. *J. Chem. Phys,* 1958, Vol. 28, N 1, 430-435.

[158] Swift, TJ; Connick, RE. NMR-relaxation of ^{17}O in aqueous solution of paramagnetic cations and the life-time of water molecules in the first coordination sphere. *J. Chem. Phys,* 1962, Vol. 37, N 2, 307-320.

[159] Voronov, VK; Ivlev, YuN; Domnina, ES; Skvortsova, GG. Assignment of resonance lines in PMR spectra of N-vinyl- and N-ethylimidazole. *Khim. Geterotsikl. Soed,* 1971, № 10. –P. 1416-1420.

[160] Voronov, VK; Ivlev, YuN; Domnina, ES; Skvortsova, GG. Paramagnetic shifts in N-vinylimidazole complexes with manganese, iron, cobalt, nickel, copper chlorides. *Khim. Geterotsikl. Soed,* 1972, № 7, 994-995.

[161] Voronov, VK; Domnina, ES; Glazkova, NP; Skvortsova, GG. NMR spectra of 1-vinyl-2-methylimidazole in the presence of Co^{2+} and Ni^{2+} paramagnetic ions. Izv. Akad. Nauk USSR. *Ser. Khim,* 1974, № 11, 2598-2599.

[162] Granet, I; Fiat, D. Effect of chemical exchange on the transverse relaxation rate of nuclei in solution containing paramagnetic ions. *J. Magn. Res,* 1974, Vol. 15, N 3, 540-548.

[163] Swift, TJ. The paramagnetic line width. NMR of paramagnetic molecules, N.Y.; L.: Acad. Press, 1973, 53-83.

[164] Carrington, A; McLachlan, AD. Magnetic resonance and its application in chemistry, M.: Mir, 1970, 447 p. (In Russian).

[165] Voronov, VK. On effect of spectrometer resonance frequency on the shifts induced by paramagnetic shifting reagents. Izv. Akad. Nauk USSR. *Ser. Khim,* 1975, № 12, 2795-2796.

[166] Obynochny, AA; Sagdeev, RZ; Molin, YuN; Rezvukhin, AI. ^{13}C NMR spectra of Ni (II) complexes with pyridines. Izv. Akad. Nauk USSR. *Ser. Khim,* 1975, № 14, 114-120.

[167] Espersen, WJ; Hutton, WC; Choio, ST; Martin, RB. Cautions concerning uses of copper(II) selective broadening in nuclear magnetic resonance spectroscopy for determination of ligand binding sites. *J. Am. Chem. Soc,* 1974, Vol. 96, N 26, 8111-8112.

[168] Espersen, WJ; Martin, RS. Predominant scalar interactions in selective broadening of ligand nuclear magnetic resonance by copper(II) ions. *J. Am. Chem. Soc,* 1976. –Vol. 98, N 1, 40-44.

[169] Beattii, JK; Fensom, DJ; Freeman, HC. Paramagnetic line broadening in the nuclear magnetic resonance spectra of amino acids and peptides. 1. The copper(II), glycine system. *J. Am. Chem. Soc,* 1976. –Vol. 98, N 2, 500-507.

[170] Peters, JA; Kieboom, APG. Multinuclear magnetic resonance in the presence of lanthanide(III) as an analytical tool for structure determination in solution. *Rec. Trav. Chim. Pays.-Bas,* 1983, Vol. 102, N 9, 381-392.

[171] Voronov, VK; Ushakov, IA. High-resolution nuclear magnetic resonance in paramagnetic complexes. *Usp. Khim,* 2010, Vol. 79, № 10, 915-928.

[172] Skvortsova, GG; Domnina, ES; Shestova, LA; et al. Interaction of indazole and alkylpyrazoles with acetylene. *Khim. Geterotsikl. Soed,* 1976, № 9, 1247-1251.

[173] Voronov, VK; VK; Shestova, LA; Domnina, ES; Skvortsova, GG. Removal of spin-spin coupling in PMR spectrum of N-vinylpyrazole by paramagnetic additives. Izv. Akad. Nauk USSR. *Ser. Khim,* 1976, № 2, 412-413.

[174] Frankel, LS. NMR investigation of relaxation and “chemical-exchange spin decoupling” in dilute solutions of Co^{2+} with various “P, O” containing solvents. *J. Chem. Phys,* 1969, Vol. 50, N 2, 943-950.

[175] Yonezawa, T; Morishima, I; Akana, Y; Fukuta, K. Nuclear magnetic resonance studies of some substituted aniline derivatives in the presence of paramagnetic Ni(II) and Co(I I) acetylacetonates. *Bull. Chem. Soc. Jap,* 1970, Vol. 43, N 2, 379-387.

[176] Voronov, VK; Moskovskaya, TE; Glukhikh, VI; et al. ^{13}C NMR study of organylhalostannane complexes with 1-vinyl- and 1-ethylimidazoles. *Coord. Khim,* 1978, Vol. 4, № 3, 388-390.

[177] Afonin, AV; Voronov, VK; Baikalova, LV; et al. Structure of 1,1-divinyl-2,2-biimidazolyl and its complex with transition metal salts. *Zh. Org. Khim,* 1987, Vol. 57, № 7, 1594-1597.

[178] Reuber, J. Paramagnetic lanthanide shift reagents in NMR spectroscopy: Principles, methodology and applications. Progress in NMR spectroscopy, 1973, Vol. 9, 1-70.

[179] Voronov, VK; Glukhikh, VI; Domnina, ES; Skvortsova, GG. Application of $(DPM)_3$ for PMR study of long-range spin-spin coupling between ^1H, ^1H and ^1H, ^{14}N.. Izv. Akad. Nauk USSR. *Ser. Khim,* 1977, № 1, 200-201.

[180] Witanowski, M; Stefanik, L; Januszewski, H; Wolkowski, Z. Selection of dipivalomethanate chelates as shift reagents for nitrogen-14 NMR. *Chem. Com,* 1971, N 23, 1573-1575.

[181] Hirayama, M; Edagava, E; Hanyu, Y. Contact term contribution to lanthanide-induced ^{13}C NMR shifts in pyridine and α -piciline. *Chem. Com,* 1972, N 24, 1342-1344.

[182] Voronov, VK; Borovik, MA; Ushakov, IA; Ivanova, NI; Malakhova, LB. NMR spectra of thiophosphoryl- and phosphorylazoles. Izv. Akad. Nauk. *Ser. Khim,* 2009, № 7, 1471-1475.

[183] Es'kova, LAVoronov, VK; Domnina, ES; Skvortsova, GG. NMR spectra and structure of 1-vinylpyrazoles. Izv. Akad. Nauk USSR. *Ser. Khim,* 1985, № 5, 1012-1018.

[184] Pelizza, Y; Nebulon, M; Ferrari, P; Gallo, G. Polymorphism of 1-methyl-2-nitro-5-vinilimidazole. *Farm. Educ. Sei,* 1978, Vol. 33, N 1, 3-13.

[185] Frolov, YuL; Chipanina. NN; Lurie, FS; et al. Intensities of IR spectra lines and intramolecular interactions in 1-vinylazoles. Izv. Akad. Nauk USSR. *Ser. Khim,* 1980, № 7, 1562-1565.

[186] Sigalov, MV; Kalabin, GA; Prpodakov, AG; et al. NMR study of conjugation effects. Izv. Akad. Nauk USSR. *Ser. Khim,* 1981, № 12, 2676-2680.

[187] Sigalov, MV; Trofimov, BA; Mikhaleva, AI; Kalabin, GA. ^1H and ^{13}C NMR study of conformational and electronic structure of 1-vinylpyrroles. *Tetrahedron*, 1981, Vol. 37, N 17, 3051-3059.

[188] Ushakov, IA; Afonin, AV; Voronov, VK; et al. ^1H and ^{13}C NMR study of spatial and electronic structures of 2-(2-acylethenyl)pyrroles. *Zh. Org. Khim,* 2003, Vol.39, № 9, 1391-1397.

[189] Knorr, R. Inductive substituent constants σ_i^i from olefinic geminal ^1H NMR coupling constants. *Tetrahedron*, 1981, Vol. 37, N 5, 929-937.

[190] Aminova, RM; Sammitov, Yu. MO LCAO calculation of ^1J, ^2J, ^3J spin-spin coupling constants dependence on orientation of lone electron pairs of nitrogen and oxygen atoms. *Zh. Strukt. Khim,* 1974, Vol. 15, № 4, 607-616.

[191] Schaefer, T; Widman, TA; Sebastian, R. The rotational angle dependence of $^5J_m^{H,SH}$ in benzenethiol. *Can. J. Chem,* Vol. 60, 1924-1927.

[192] Voronov, VK; Es'kova, LA; Domnina, ES; Skvortsova, GG. Longe-range spin-spin coupling and relative orientation of vinyl group in 1-vinylpyrazoles. Izv. Akad. Nauk USSR. *Ser. Khim,* 1981, № 10, 2363-2365.

[193] Afonin, AV; Garashchenko, ZV; Voronov, VK; et al. ^1H and ^{13}C NMR study of vinyloxyphenylazomethines structure. *Zh. Org. Khim,* 1988, Vol. 24, № 6, 1284-1287.

[194] Voronov, VK; Pogodayeva, NN; Tyukavkina, NA, Analysis of PMR spectra of 3- and 5-substituted flavons with the paramagnetic additives. *Khim. Geterotsikl. Soed,* 1974, № 11, 1468-1471.

[195] Wuersberg, E. Temperature effect on the absorption spectrum of europium(3+) in solution. *J. Chem. Phys,* 1976, Vol. 64, N 3, 1254-1256.

[196] Voronov, VK. Contact interaction and relaxation effects in paramagnetic complex of $Eu(FOD)_3$ complex with 3- and 5-oxyflavon. Izv. Akad. Nauk USSR. *Ser. Khim,* 1977, № 8, 1878-1880.

[197] Frolov, YuL; Sapozhnikov, YuM; Chipanina, NN; et al. On correlations of pK values in basic and electron-excited states. Izv. Akad. Nauk USSR. *Ser. Khim,* 1978, № 2, 301-304.

[198] Buchachenko, AL. Chemical polarization of electrons and nuclei, M.: Nauka, 1974, 246 p.

[199] Rackam, DM; Cockerill, AF; Harrison, RG. Proton and carbon magnetic resonance study of specific contact interactions of a bifunctional oxazole amide with jluorinated lanthanide shift. *Org. Magn. Res,* 1978, Vol. 11, N 8, 424-428.

[200] Voronov, VK; Andriyankov, MA; Skvortsova, GG. Application of paramagnetic additives in structural researches. Vinyl ethers of pyridylcarbinols. *Khim. Geterotsikl. Soed,* 1975, № 5, 666-669.

[201] Babailov, SP; Mainichev, DA. NMR studies of mixed-ligand lanthanide (III) complexes. Peculiarities of molecular structure, dynamics and paramagnetic properties for cerium subgroup chelates with crown ethers. *J. Inc. Phenom. Macrocyclic Chem,* 2002, Vol. 43, 187-193.

[202] Kalabin, GA; Trofimov, BA; Amosova, SV; Voronov, VK. NMR analysis of 2,5-dimethyl-4-methylene-1,3-oxathiolane structure. *Khim. Geterotsikl. Soed,* 1975, № 5, 606-611.

[203] Sigalov, MV; Voronov, VK; Andriyankov, MA; Skvortsova, GG. Aroxydihydropyranes. XII. Configuration and conformation of 8-(3,4-dihydropyranyl-2)oxyquinoline derivatives. *Khim. Geterotsikl. Soed,* 1976, № 5, 607-610.

[204] Voronov, VK; Trofimov, BA; Gavrilova, GM; et al. PMR spectra and structure of 2H,6H-2,6-dimethyl-4-amino-1,3,5-dithiazine and its derivatives. *Khim. Geterotsikl. Soed,* 1981, № 4, 485-487.

[205] Demarco, PV; Elzey, TK; Lewis, RB; Wenkert, E. Paramagnetic induced shifts in the proton magnetic resonance spectra of alcohols using tris (dipivalomethanato)europium(I I I). *J. Am. Chem. Soc,* 1970, Vol. 92, N 19, 5734-5737.

[206] Demarco, PV; Elzey, TK. Tris(dipivalomethanato)europium(I I I). A shift reagent for use the proton magnetic resonance analysis of steroids and terpenoids. *J. Am. Chem. Soc,* 1970, Vol. 92, N 19, 5737– 5739.

[207] Buckley, DC; Green, GH; Ritchie, GH; Taylor, WC. Application of the tris(dipivalomethanato)europium nuclear magnetic resonance shift reagent the study of triterpenoids. *Chem. And Ind,* 1971, N 11, 298-299.

[208] Briggs, I; Frost, GH; Hart, FA; et al. Lanthanide-induced shifts in nuclear magnetic resonance spectroscopy. Shifts to high field. *J. Chem. Soc. Chem. Com,* 1970, N 12, 749-750.

[209] Briggs, I; Hart, FA; Moss, GP. The application of lanthanide-induced shift to the complete analysis of the borneol nuclear magnetic resonance spectrum. *J. Chem. Soc. Chem. Com,* 1970, N 22, P.1506-1507.

[210] Willcott, MR; Oth, IFM; Thio, I; et al. Rapid solution of stereochemical problems with europium-tris(tetramethylheptanedione), $Eu(thd)_3$. *Tetrahedron Lett,* 1971, N 19, 1579-1582.

[211] Tjan, SB; Visser, FR. PMR shift to high field induced by tris (dipivalomethanato)europium. *Tetrahedron Lett,* 1971, N 30, 2833-2836.

[212] Beate, C; Wolkowski, ZW; Thoai, N. Tris(dipivalomethanato)ytterbium-induced shifts in PMR amines. *Tetrahedron Lett,* 1971, N 12, 817-820.

[213] Wolkowski, ZW, Tris(dipivalomethanato)ytterbium-induced shifts in PMR spectra of ketones and aldehydes. *Tetrahedron Lett.* 1971, N 12, 821-824.

[214] Okutani, T; Morimoto, A; Kenekc, T; Masuda, K. Studies on azetidine derivatives. III. Configuration assignment with a shift reagent. *Tetrahedron Lett.* 1971, N 16, 1115-1118.

[215] Hinckley, CC; Klotz, MR; Patil, F. Applications of rare earth nuclear magnetic shift reagents. III. Graphical analysis of paramagnetic shifts for systems having two coordination sites. Testosterone and 17-methyltestosterone. *J. Am. Chem. Soc,* 1971, Vol. 93, N 10, 2417-2420.

[216] Fleming, I; Hanson, SW; Sanders, JKM. The effect of $Eu(DPM)_3$ on the NMR spectra of bifunctional compounds. *Tetrahedron Lett,* 1971, N 40, 3733-3736.

[217] Morallee, KG; Nieboer, E; Rossotti, FJC; et al. The lanthanide cations as nuclear magnetic resonance probes of biclogical systems. *J. Chem. Soc. Chem. Com,* 1970, N 18, 1132-1133.

[218] Crump, DR; Sanders, JKM; Williams, DH. Evaluation of some tris (dipivalomethanato)lanthanide complexes as paramagnetic shift reagents. *Tetrahedron Lett,* 1971, N 50, 4419-4422.

[219] Horrocks, W DeW; Sipe, JP. Lanthanide shift reagents. A Survey. *J. Am. Chem. Soc,* 1971, Vol. 93, N 25, 6800-6804.

[220] Whitesides, GM; Lewis, DW. Tris(tret-butylhydroxymethylene)-*d*-camphorato europium(III). A reagent for determining enanthiomeric purity. *J. Am. Chem. Soc,* 1970, Vol. 92, N 23, 6979-6980.

[221] Fraser, R; Petit, MA; Sanders, JKM. Determination of enanthiomeric purity by an optically active nuclear magnetic resonance shift reagent of wide applicability. J. Che. Soc. *Chem. Com,* 1971, N 22, 1450-1451.

[222] Creene, JL; Shevlin, PB. Meso- and (±)-bis(phenysulphinyl) methane. Characterization using nuclear magnetic resonance chemical shift reagents. J/ Chem. Soc. *Chem. Com,* 1971, N 18, 1092-1093.

[223] Witanowski, M; Stefaniak, L; Januszewski, H. Effect of $Yb(DPM)_3$ and $Eu(DPM)_3$ on nitrogen-14 magnetic resonance. *Tetrahedron Lett,* 1971, N 20, 1653-1656.

[224] Briggs, J; Hart, FA; Moss, GP; Randall, EW. A real method of assignment for ^{13}C spectrum of borneok. *J. Chem. Soc. Chem. Com,* 1971, N 7, 304-305.

[225] Birdsall, B; Feenly, J; Glasel, JA; et al. A method of assigning ^{13}C nuclear magnetic resonance spectra using europium(III) ion-induced pseudocontact shifts and C–H heteronuclear spin decoupling techniques. Chem. Soc. *Chem. Com,* 1971, N 22, 1473-1474.

[226] Sanders, JKM; Williams, DH. Evidence for contact and pseudo-contact contributions in lanthanide-induced ^{31}P NMR shifts. *Tetrahedron Lett,* 1971, N 30, 2813-2816.

[227] Hinckley, CC; Boyd, WA; Smith, GV. Deuterium isotope effects observed in NMR shifts induced by lanthanide complexes. *Tetrahedron Lett,* 1972, N 10, 879-882.

[228] Grotens, AM; Hilbers, CW; Beer, E. Isotope effects in nuclear magnetic resonance spectra of polyglycoloimethylethers, complexed with rare-earth shift reagents. *Tetrahedron Lett,* 1972, N 20, 2067-2070.

[229] Willcott, MR; Lenkinski, RE; Davis, RE. Interpretation of the pseudocontact model for nuclear magnetic resonance shift reagents. I. The agreement factor R. *J. Am. Chem. Soc,* 1972, Vol. 94, N 5, 1742-1744.

[230] Davis, RE; Willcott, MR. Interpretation of the pseudocontact model for nuclear magnetic resonance shift reagents. II. Significance testing on the agreement factor *R. J. Am. Chem. Soc,* 1972, Vol. 94, N 5, 1744-1745.

[231] Lefevre, F; Martin, ML. Use of rare-earth metal complexes in nuclear magnetic resonance. *Org. Magn. Res,* 1972, Vol. 26, N. 4(5), 737-763.

[232] Willcott, MR; Davis, RE; Faulner, DJ. The configuration and conformation of 7-chloro-1,6-dibromo-3,7-dimethyl-3,4-epoxy-1- octene. *Tetrahedron Lett,* 1973, N 40, 3967-3970.

[233] Latkovskaya, SV; Bespalov, VI; Korobitsyna, IK. Configuration of oximes of some tetra-substituted five-membered cyclic ketones. *Zh. Obshch. Khim*, 1974, Vol. 10, № 11, 2336-2342.

[234] Brooke, GM; Matthew, RS. The use of a lanthanide shift reagent in ^{19}F NMR spectroscopy. The orientation of fluorine atoms in a trifluoroidanol. *Tetrahedron Lett,* 1973, N 36, 3469-3470.

[235] Cheer, CJ; Rosen, W; Webel, JJ. Lanthanide induced shift NMR studies of some bicycle-6,10-nones-2,4,6-triene derivatives. *Tetrahedron Lett,* 1974, N 46, 4045-4048.

[236] Williamson, KL; Clutter, DR; Emoh; et al. Conformational analysis by nuclear magnetic resonance shift reagent studies on acyclic alcohols. ^{1}H and ^{13}C spectra of the six-carbon aliphatic alcohols. *J. Am. Chem. Soc,* 1974, Vol. 96, N 5, 1471-1479.

[237] Quereshi, MS; Walker, JM. Nuclear magnetic resonance studies on the structures of some substituted N-alkylpyridinium pentanitratolanthanide ion pairs in solution. *Inorg. Chem,* 1974, Vol. 13, N 12, 2896-2900.

[238] Jnagaki, F; Takanashi, S; Tasunu, M; Mijazawa, T. Conformation of thymidine 3,5′-cyclic phosphate in aqueous solution. Compatibility of the information from the lanthanide ion probe and the spin-spin coupling constants. *Bull. Chem. Soc. Jap,* 1976, Vol. 49, N 3, 611-616.

[239] Flochhart, BO. Lanthanide shift reagents in nuclear magnetic resonance spectroscopy. Crit. Revs. *Anal. Chem,* 1976, Vol. 6, N 1.– 69-130.

[240] Mayo, BC. Lanthanide shift reagents in NMR spectroscopy. *Chem. Soc. Rev,* 1973, Vol. 2, N 1, 49-74.

[241] Cockerill, AF; Davis, GLO; Harden, RC; Rackham, DM. Lanthanide shift reagents for NMR spectroscopy. *Chem. Rev,* 1973, Vol. 73, N 6, 553-588.

[242] Slonim, IYa; Bulay, AKh. Paramagnetic shifting reagents in NMR spectroscopy. *Usp. Khim,* 1973, Vol. 42, № 11, 1976-2006.

[243] Voronov, VK. Paramagnetic reagents for studying the structure of organic ligands. *Usp. Khim,* 1974, Vol. 43, № 3, 432-454.

[244] Kornilov, MYu; Turov, AV. Lanthanide shifting reagents in chemistry of heterocyclic compounds. *Khim. Geterotsikl. Soed,* 1979, № 10, 1299-1316.

[245] Butkus, EP; Kadzyauskas, PP. Conformational study using NMR method with lanthanide shifting reagents. *Zh. Strukt. Khim,* 1985, Vol. 26, № 2, 145-166.

[246] Babailov, SP; Krieger, YuG. NMR spectroscopic methods of investigations of molecular structure of paramagnetic complex compounds of lanthanides in solutions and their application for complexes with crown ethers. *Zh. Strukt. Khim,* 1998, Vol. 39, № 4, C. 714-730.

[247] Morii, K; Maruta, G; Takeda, S. Solid-state high-resolution NMR studies on spin fluctuation associated with valence tautomerism of metal-benzoquinone system: Cobalt complex in the stable and metastable phases. *Polyhedron*, 2005.– Vol. 24, N 16-17, 2607-2613.

[248] Babailov, SP. Lanthanide paramagnetic probes for NMR spectroscopic studies of molecular conformational dynamics. Applications to macrocyclic molecules. *Progr. Nucl. Magn. Reson. Spectrosc*, 2008, Vol. 52, N 1, 1-21.

[249] Abraham, A; Bleney, B. Electron paramagnetic resonance of transition ions, *M.: Mir*, 1973, Vol. 1, 661 p.

[250] Rondeau, RE; Sievers, RE. New nuclear magnetic resonance shift reagents. *Anal. Chem,* 1973, Vol. 45, N 12, 2145-2147.

[251] Horrocks, W DeW. Water-soluble lanthanide porphyrins: shift reagents for aqueous solution. *J. Am. Chem. Soc,* 1978, Vol. 100, N 14, 4386-4392.

[252] Burns, PD; La Mar, GN. Proton spin relaxation for the nonlabile reagent. *J. Magn. Res,* 1982, Vol. 46, N 1, 61-68.

[253] Levy, GC; Edlund, U; Holloway, CE. Mechanisms for interactions between organic molecules and paramagnetic relaxation reagents. *J. Magn. Res,* 1976, Vol. 24, N 3, 375-387.

[254] Levy, GC; Dechter, JJ; Kowalewski, J. The effect of paramagnetic relaxation reagents on ^{15}N spin relaxation and the use of $Gd(dpm)_3$ as a nitrogen-15 nuclear magnetic resonance spin label. *J. Am. Chem. Soc,* Vol. 100, N 8, 2308-2314.

[255] Zelonka, RA; Baird, MC. Spin exchange between paramagnetic metal complexes and aliphatic nitroxides: fluorine-19 nuclear magnetic resonance spectroscopy of copper (II) complexes. *J. Am. Chem. Soc,* 1970, N 21, 1448-1451.

[256] Hinckley, CC; Brumley, WC. Effects of random coordinate error in analyses of lanthanide-induced pseudocontact shifts. Axially symmetric case. *J. Am. Chem. Soc,* 1976, Vol. 98, N 6, 1331-1336.

[257] Hinckley, CC; Brumley, WC. Errors in analyses of lanthanide-induced shifts. Cis- and trans-pinocarved. *J. Magn. Res,* 1976, Vol. 24, N 2, 239-250.

[258] Richardson, MF; Rothstein, SM; Li, WK. Significance testing of lanthanide shift reagent data. *J. Magn. Res,* 1979, Vol. 36, N 1, 69-70.

[259] Keiko, VV; Voronov, VK. Determination of mutual orientation of ligand molecules and coordination ion in paramagnetic. Izv. Akad. Nauk USSR. *Ser. Khim,* 1976, № 6, 125-126.

[260] Kornilov, MYu; Turov, AV; Ishchenko, AA; Tolmachev, AI. On geometrical structure of adducts of lanthanide shift reagents with carbonyl compounds. *Theor. Experiment. Khim,* 1975, Vol. 11. № 5, 690-695.

[261] Danovich, DK; Zakzhevsky, VG; Voronov, VK, Software package for computation of spatial structure of molecules using a method of paramagnetic additives. *Zh. Strukt. Khim,* 1986, Vol. 27, № 2, 165-166.

[262] Zolin, VF; Koreneva, LG. Application of the mathematical modeling of electronic spectra for structural studies of lanthanide complexes in solution. *Zh. Strukt. Khim,* 1983, Vol. 24, № 5, 60-66.

[263] Reuben, J. Lanthanides as spectroscopic and magnetic resonance probes in biological systems. *Naturwiss*, 1975, Vol. 62, N 4, 172-178.

[264] Reuben, J. Structural information from chemical shift in lanthanide complexes. *J. Magn. Res,* 1982, Vol. 50, N 2, 233-236.

[265] Zijl, PCM; Wezel, RP; Maclean, C; Bothider-By, AA. Evaluation of the dipolar contribution to lanthanide shift reagent induced isotropic shifts. *J. Phys. Chem*, 1985, Vol. 89, N 2, 204-207.

[266] Reuben, J; Leigh, JS. Effects of paramagnetic lanthanide shift reagents on the proton magnetic resonance spectra of quinoline and and pyridine. *J. Am. Chem. Soc,* 1972, Vol. 94, N 8, 2789-2793.

[267] Zaev, EE; Molin, YuN. NMR paramagnetic shifts and structure of Co(II) and Ni(II) acetylacetonate complexes with Npyridine oxide. *Zh. Strukt. Khim,* 1966. Vol. 7, № 5, 680-686.

[268] Monduzzi, M; Lai, A; Saba, C; et al. C-13 magnetic relaxation rates and H-1 and C-13 paramagnetic shifts of Co(I I) complex of dopamine. *Adv. Mol. Relaxation int. Processes*, 1982, Vol. 24, N 4, 233-244.

[269] Voronov, VK; Kalinovskaya, GP; Enikeeva, EI. On HFC nature in paramagnetic complexes of rare-earth elements with organic molecules. *Zh. Obshch. Khim*, 1984, Vol. 54, № 9, 2159-2160.

[270] Daman, ME; Dill, K. Use of electron-nuclear relaxation-rates to detetmine Mn^{2+}-metyl-*D*-galactopyranoside interactions. *Carbohydr. Res*, 1984, Vol. 132, N 2, 335-338.

[271] Reference book of chemist: 2 Vol, 3-rd ed. / Ed. B.P. Nikolsky, L.: Khimiya, 1871, Vol. 1, 303 p.

[272] Voronov, VK; Shagun, VA; Keiko, VV; et al. On structure of 1-vinylimidazole and possible inversion of the pyrrole nitrogen. *Zh. Strukt. Khim,* 1980, Vol. 21, № 5, 71-75.

[273] Cramer, RE; Seff, K. Crystal and molecular structure of NMR shift reagent, $Eu(dpm)_3(py)_2$. *Chem. Com,* 1972, N 7, 400-401.

[274] Gansow, OA; Loeffler, PA; Davis, RE; et al. Contact vs. pseudocontact contributions to lanthanide-induced shifts in the NMR spectra of isoquinoline and of endo-norbornenol. *J. Fm. Chem. Soc*, 1976, Vol. 98, N 14, 4250-4258.

[275] Gillespie, R. Geometry of molecules, M.: Mir, 1985, 278 p. (In Russian).

[276] Kashayev, AA; Zelsbt, EA; Demidov, MP; et al. Crystalline and molecular structure of 1-vinylimidazole complexes with copper and cobalt dichlorides. *Coord. Khim,* 1978, Vol. 4, № 5, 785-790.

[277] Sokol, VI; Zefirov, YuV; Porai-Koshits, MA. Crystalline and molecular structure of dichloro-bis (vinylbenztriazole) cobalt (II). *Coord. Khim,* 1979, Vol. 4, № 8, 1249-1258.

[278] Morrison, R; Boyd, R. Organic Chemistry, M.: Mir, 1974, 1132 p. (In Russian).

[279] Torck, F; Negebus, A; Pulay, P. Calculation of fully optimised geometries of five- and six-membered heterocycles by the CNDO force method. *Teor. Chim. Acta*, 1973, Vol. 32, N 2, 146-150.

[280] Richards, WG; Wallis, J. Imidazole geometry and the calculation of tautomer preference for histamine. *Eur. J. Med. Chem*, 1979, Vol. 12, N 1, 9-12.

[281] Skelton, BW; White, AH. Crystal structure of bromo(tetraphenylporphyrin)iron(I I I). *Aust. J. Chem,* 1977, Vol. 30, 2655-2660.

[282] Voronkov, MG; Dyakov, VM. Silatranes, *Novosibirsk: Nauka*, 1978, 206 p.

[283] Voronov, VK; Keiko, VV; Baryshok, VP; et al. Coordination of 1-methylsilatrane with lanthanide shifting reagents. Dokl. Akad. Nauk. USSSR. *Ser. Khim,* 1977, Vol. 147, № 1, 147-149.

[284] Voronov, VK; Voronkov, MG; Dyakov, VM; Baryshok, VP, Proton magnetic resonance of some Si-substituted silatranes in the presence of Ni^{2+} and Co^{2+}paramagnetic ions. Dokl. Akad. Nauk. USSSR. *Ser. Khim,* 1978, № 6, 1457-1459.

[285] Voronov, VK; Voronkov, MG. The complexing ability and structure of Si-substituted silatranes. *J. Mol. Struct,* 1980, Vol. 67, 285-287.

[286] Yonezawa, T; Morishima, I; Ohmori, Y. Studies on the proton paramagnetic shifts for σ -systems in piperidine and quinuclidine. *J. Am. Chem. Soc,* 1970, Vol. 93, N 5, 1267-1274.

[287] Clardy, JC; Milbrath, DS; Verkade, JG. Unusual phosphorus basicity of $P(OCH_2CH_2)_3N$ and the molecular structure of $H_3BP(OCH_2CH_2)_3N$. *Inorg. Chem,* 1977, Vol. 16, N 8, 2135-2137.

[288] Paddon-Row, MN; Radow, L; Grerory, AR. 1–aza-4-boretricyclo[2.2.2.0^{1,4}]octane and 1–aza-4-borabicyclo[2.2.2]octane: stable bond-strech isomers. An ab initio study. *Chem. Com,* 1976, N 12, 427-428.

[289] Stohrer, W-D; Hoffman, R. The electronic structure and reactivity of strained tricyclic hydrocarbons. *J. Am. Chem. Soc.* –1972, Vol. 94, N 3, 779-786.

[290] Soldatenko, AS. Synthesis of new intra - and intermolecular complex compounds of hypervalent silicon: Ph. D. Thesis, *Иркутск*, 2011, 160 p.

[291] Trofimov, BA. Heteroatom derivatives of acetylene, M.: Nauka, 1981, 319 p.

[292] Voronov, VK; Nesterenko, RN; Mikhaleva, AI; Kanitskaya, LV, Structural study of methyl thienyl ketone oxime using a method of paramagnetic additives. *Zh. Strukt. Khim,* 1986, Vol. 27, № 2, 163-165.

[293] Krivdin, LB; Larina, LI; Chernyshev, KA; Rozentsveig, IB. Non-empirical calculations of NMR indirect spin-spin coupling constants. Part 13 - Configurational assignment of aminosulfonylamidines. *Magn. Reson. Chem*, 2005, V. 43, N. 11, 937-942.

[294] Krivdin, LB; Larina, LI; Chernyshev, KA; Rulev, AYu. Non-empirical calculations of NMR indirect spin-spin coupling constants. Part 14. Azomethines of the α,β-unsaturated aldehydes. *Magn. Reson. Chem*, 2006, V. 44, N. 2, 178-187.

[295] Sagdeev, RZ; Voronov, VK; Podoplelov, AV; et al. Peculiarities of NMR spectra of heterospin complexes. Izv. Akad. Nauk. *Ser. Khim,* 2001, № 11, 1986 –1994.

[296] Drago, R. Physical methods in chemistry, M.: Mir, 1981, Vol. 2-452 p.

[297] Ovcharenko, VI; Romanenko, GV; Ikorskii, VN; et al. Polymorphous modifications of a Ni^{2+}| complex with stable nitroxide involving Ni^{2+}-O.bul.-N bonds. Quantum-chemical investigation of exchange imteractions in heterospin systems. *Inorg. Chem,* 1994, Vol. 33, N 15, 3370 –3381.

[298] Sato, O; Tao, J; Zhang, Y-Z. Control of magnetic properties through external stimuli. *Angew. Chem. Int. Ed*, 2007, Vol. 46, 2152-2187.

[299] Voronov, VK; Ushakov, IA; Shmelev, VV; Sagdeev, DP. Peculiarities of intramolecular exchange and valence tautomerism in metal semiquinolates determined by high-resolution NMR spectroscopy. *Magn. Reson. Chem*, 2012.–Vol. 50, N 5, 350–356.

INDEX

A

B

C

D

E

F

G

H

I

T

U

V

W

X

Y

Z